Weather Forecasting
A Young Meteorologist's Guide

Weather Forecasting
A Young Meteorologist's Guide

Dan Ramsey

TAB BOOKS
Blue Ridge Summit, PA

FIRST EDITION
SECOND PRINTING

© 1990 by **TAB Books**.
TAB Books is a division of McGraw-Hill, Inc.

Printed in the United States of America. All rights reserved. The publisher takes no responsibility for the use of any of the materials or methods described in this book, nor for the products thereof.

Library of Congress Cataloging-in-Publication Data

Ramsey, Dan, 1945-
    Weather forecasting : a young meteorologist's guide / by Dan Ramsey.
      p.   cm.
    Summary: Discusses such aspects of weather forecasting as the elements involved in causing weather conditions, basic and advanced instruments, and the collecting and interpreting of data.
   ISBN 0-8306-8338-0   ISBN 0-8306-3338-3 (pbk.)
   1. Weather forecasting—Juvenile literature. [1. Weather forecasting.] I. Title.
QC995.43.R36   1990
551.6′3—dc20   89-48463
                                                                           CIP
                                                                           AC

TAB Books offers software for sale. For information and a catalog, please contact TAB Software Department, Blue Ridge Summit, PA 17294-0850.

Acquisitions Editor: Kimberly Tabor
Book Editor: Michael R. Porcellino

This book is dedicated to Heather Kay
Who makes the Sun shine on a cloudy day.

Contents

Acknowledgments — xi

Introduction — xiii

1 The Atmosphere — 1
Mankind vs the Elements *1*
The Climate *2*
The Sun *2*
The Earth *5*
Solar Radiation *8*
Layers of the Atmosphere *9*
Climate Controls *13*
Do You Know . . . *16*

2 Air Masses — 19
Air Mass Classifications *19*
Source Regions *21*
How Air Masses Change *23*
Understanding Fronts *26*
Cold Fronts *27*
Warm Fronts *29*
Occluded Fronts *30*
Stationary Fronts *32*
Lows and Highs *35*
Do You Know . . . *38*

3 The Elements — 39
Temperature *39*

Causes of Temperature Change *41*
Pressure *45*
Lows, Highs, and Pressure Gradients *49*
Wind *50*
Moisture *53*
Clouds *58*
The Hydrologic Cycle *61*
Do You Know . . . *61*

4 The Weather Eye — 67
The National Weather Service *68*
Weather Proverbs *68*
The Human Condition *72*
Weather and the Mind *73*
Sky Observations *73*
Keeping Track *74*
Do You Know . . . *75*

5 Basic Instruments — 77
The Weather Shack *78*
Wind Instruments *78*
Barometer *80*
Thermometers *83*
Hygrometers *85*
Rain Gauges *89*
Measuring Clouds *90*
Do You Know . . . *92*

6 Collecting Weather Data — 93
The Weather Log *93*
Collecting Data from the Weather Service *99*
Distribution of Weather Information *101*
Reading the Weather Map *106*
The Station Model *107*
Do You Know . . . *109*

7 Interpreting Weather Data — 113
Maritime Polar Air Masses *113*
Continental Polar Air Masses *115*
Maritime Tropical Air Masses *116*
Continental Tropical Air Masses *117*
Plotting the Weather *117*
Forecasting Your Own Weather *118*
How the NWS Forecasts Weather *121*
Climatology *123*
Considering Micro-Climates *123*

Improving Forecast Accuracy   *125*
  Do You Know . . .   *126*

8 Using The Weather   **127**
  Agrometerology   *127*
  The Farm Weather Forecaster   *128*
  Weather and the Pilot   *130*
  Weather and Boaters   *134*
  Industrial Meteorology   *136*
  Weather and Comfort   *136*
  Wind Chill   *139*
  Using Daily Weather   *139*
  Do You Know . . .   *139*

Glossary   **141**

Index   **145**

Acknowledgments

Many people and groups helped in the research and writing of this book. They should be thanked for making the complex science of meteorology into the simple hobby of weather-watching. They include: U.S. Department of Commerce, National Oceanic and Atmospheric Administration, National Weather Service, Environmental Data Service, U.S. Department of Agriculture's Forest Service, Federal Aviation Administration, Department of the Army, the Naval Education and Training Support Command, and the Heath Company of Benton Harbor, MI.

Individual thanks go to Donald E. Witten and Lucinda Thorpe with NOAA's National Weather Service and Rob Boatman of Heath Company. Heather Ramsey earns a special thank you for her help in typing the manuscript and preparing illustrations.

Introduction

Benjamin Franklin once said, "Some people are weather-wise, but most are otherwise."

Poor Richard was right, yet no one will deny the importance of weather to our daily lives. Weather brings snow to play in, rain to grow crops, and sunny summer days. Weather can produce economic prosperity or disaster. Weather affects business, government, and education. It impacts every day of our lives. We can eliminate the unfavorable effects—and even profit from them—with a simple knowledge of what weather is, what it does, and how to forecast and use it.

Weather Forecasting: A Young Meteorologist's Guide is a simple and entertaining book on how you can be an amateur meteorologist for fun and profit—even if you can't pronounce meteorologist (mee–tee–or–ALL–o–jist). The first three chapters explain what weather is, where it comes from, and what makes summer hot and winter cold, fog lift, and rain fall. Then you'll learn about the *air masses* and *fronts* that the TV weatherforecasters always point to on maps. And you'll consider each of the individual elements that are collectively called *the weather:* wind, barometric pressure, temperature, humidity, precipitation (rain, snow, etc.) and clouds.

Once you understand the basics of weather, you'll learn to observe weather and how it moves in and out of your life each day. Weather folklore and fairy tales are considered in chapter 4—some of them useful and some humorous. Observation gets more scientific as you learn in chapter 5 how to make and choose your own weather shack instruments for measuring the elements.

You will learn in chapter 6 how to collect and record the type of weather data needed to make accurate forecasts of weather 1, 2, 5, and even 90 days in advance. In Chapter 7 the data is interpreted using methods as simple or as complex as you wish—from a barometer and a chart to a numerical weather computer. You'll even find out how to let the National Weather Service do all the hard work while you improve on their forecasts and make them more accurate for your location.

Chapter 8 details how weather is used every day by all kinds of people—students, farmers, gardeners, aviators, sports stars, travelers, ships' captains, pleasure boaters, business operators, recreation directors, salespeople, army generals, housewives, and blue collar workers. You'll learn what comfortable weather really is, and you'll find out how to put the weather to work for you.

Along the way you'll discover weather through simple projects and experiments. You'll even learn how to build your own weather instruments, construct you personal weather station, and record weather information.

Each chapter offers a list of review questions to help ensure that you understand the weather. There's also a comprehensive glossary of weather words at the end of this book.

Most important, more than 100 charts, diagrams, photographs, and drawings are included to make the weather easy for you to understand and predict.

1
The Atmosphere

WEATHER IS THE CONDITION OF THE ATMOSPHERE AROUND US MEASURED IN TERMS OF HEAT, pressure, wind, and moisture. It's a warm, sunny day with light winds. It's a blizzard or a thunderstorm. It's cloudy skies and rain.

Meteorology is the science that considers—and tries to do something about—the atmosphere and its elements. It is a physical science that looks at the past and present conditions of the to predict or *forecast* the weather in the future.

Meteorology can be as simple or as complex as man's understanding. Satellites and computers are used to remind us of what the mariner already knows: "Red sky in morning, sailors take warning. Red sky at night, sailors delight."

With complexity comes greater accuracy, usually, as man's ability to foresee weather conditions is extended. A century ago mankind knew little more about the what, why, and when of weather than he did a thousand years before. The past few decades have made it easier for him to accurately predict weather days and even weeks in advance.

And mankind has learned to profit from his predictions.

MANKIND VS. THE ELEMENTS

Not that long ago, mankind was nearly always the victim, never the victor, of the elements. When it rained, he got wet. When the sun shone, he was tanned. The weather decided much of what he would eat. The passage of time was counted by "rainy seasons" and "droughts."

Then he learned that the branches of trees offered shelter from the direct rays of the Sun and caves kept the rain off his head. Trees and caves were difficult to move so he had a bright idea: movable shelters of wood and rock. He soon learned that watching the horizon would give him advance warning of rain and winds. He also discovered that animals acted differently just before a storm. He could then foretell, or *forecast*, the weather and take advantage of the elements to improve his life. The weather became his tool.

But his knowledge of current and future conditions was limited to what he could see. He was still a victim of any weather conditions beyond the horizon. What he needed was a friend who lived over the hill and was willing to run over a few times a day to tell him what weather was coming. It was not until the invention of the telegraph that he could extend his weather observations beyond a few miles and a few hours.

Since that time, mankind has developed the art of weather forecasting to a science. It is a science that, in all levels of society, uses the natural elements as tools. The farmer can use the weather to decide if he should grow melons or mint. Mothers can know weather to dress their children in swimming trunks or snowsuits. Merchants can plan sidewalk sales. Families can plan picnics. Weather surprises still occur, but with less frequency than the days when mankind knew little about the weather.

THE CLIMATE

Climate is weather plus time. That is, the climate of an area is the type and amount of weather elements that occur over a long period of time. It might be sunny today in Seattle, Washington, but it's climate is rainy. Climate is the history of weather in a location and includes averages, extremes, and trends in temperature, precipitation, humidity, wind, cloudiness, and snow cover (FIG. 1-1).

Climatology (cli-ma-TOL-o-gee) is the scientific study of climate. A climatologist may study the causes of climatic data for agriculture, industry, aviation, or any of many fields.

Weather—thus climate—is caused by the interaction of the Sun and the Earth. If for some reason the Sun were to brighten or diminish or the Earth were to tilt or move from its place in space, our weather would change and life as we know it might soon come to a screeching halt.

In order to understand weather, let's take a simplified look at the Sun, the Earth, its atmosphere, and the way they interact to give us warm days and growing seasons.

THE SUN

The Sun is Earth's only source of heat energy and the cause of all weather and atmospheric changes on Earth. With a surface temperature of about 11,000 degrees Fahrenheit (°F), the Sun radiates energy in all directions. The Earth intercepts about one two-billionths of the energy radiated by the Sun in the form of light waves. Even so, better than 99.9 percent of the Earth's heat comes from the Sun.

The Sun is a globe of gas heated by thermonuclear reactions from within the

1-1 Generalized physical regions in the United States.

central core (FIG. 1-2). The tremendous heat energy generated within the Sun's core is conveyed by the transfer of photons, which bounce from atom to atom, much like bouncing balls. Within the convective zone, which extends very nearly to the Sun's surface, the heated gases are raised upward with some cooling and subsequent convective action occurring until the gases are cooled to about 600°K (Kelvin or Absolute) at the Sun's surface.

The main body of the Sun, although composed of gases, has a well-defined, visible surface referred to as the *photosphere*. This is what we see if we were to look at the Sun (which you should never do). From the photosphere all the light and heat of the Sun are radiated. Above the photosphere is a clear gasous layer called the *chromosphere* with a thickness of about 6000 miles. Above the chromosphere is the *corona*, which may extend outward a distance of several solar diameters. As you can see in FIG. 1-2, the photosphere, chromosphere, and corona make up the solar atmosphere.

Within the solar atmosphere, solar activity occurs which affects our weather. *Solar prominences* are perhaps the most beautifully colored parts of the sun. They appear as great clouds of gas, sometimes resting on the Sun's surface, at other times floating free with no visible connection. When viewed against

4   THE ATMOSPHERE

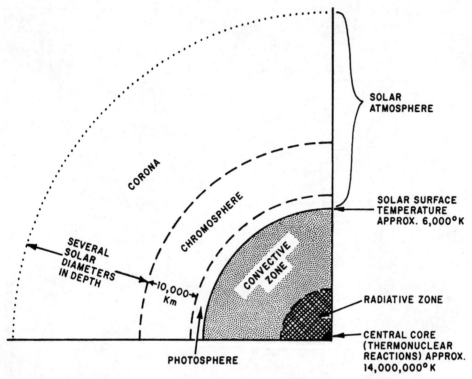

1-2   One-quarter cross section of the Sun's structure.

the solar disc they appear as long dark filaments. They display a variety of shapes, sizes, and activity that are hard to describe. The more active types appear hotter than the surrounding atmosphere with temperatures near 17,000°F.

Sunspots are dark areas on the surface of the sun. They may appear singly or in groups with large spots near the center. Sunspots begin as small dark areas known as *pores*. These pores develop into full-fledged spots in a few days with maximum development occurring in about one to two weeds. Sunspot decay consists of the spot shrinking in size. This life cycle may run from a few days for small spots to nearly 100 days for larger groups. The larger spots are about 75,000 miles across. Sunspots appear to come and go in cycles of about 8 to 17 years.

Plages are large, irregular, bright patches that surround sunspot groups. They normally appear with solar prominences and may be in round or spiral patterns. Plages are features of the lower chromosphere and are often blocked by an underlying sunspot.

Solar flares are perhaps the most spectacular of solar activity. They appear as flecks of light that suddenly appear near activity centers, appearing quickly as though a switch were thrown. They rise sharply to peak brightness in a few minutes, then decline gradually. The number of flares may increase rapidly over an area of activity. Small flare-like brightenings are always in progress during the more active times of activity.

The smaller flares can be classified as subflares. In some cases, flares may take the form of prominences, violently spewing material into the solar atmosphere and breaking into smaller high-speed blobs or clots. Flare activity varies widely between solar activity centers and seems to be tied to the magnetic field. The greatest flares occur during the week or ten days when sunspot activity is at its maximum.

THE EARTH

Of the nine planets of our solar system, the Earth is the third from the Sun. Its maximum distance from the Sun is 94 million miles in summer. Its minimum distance from the Sun is 91 million miles in winter. It has an atmosphere more than 60 miles thick.

The Earth is subject to four motions in its movement through space. Only two of these motions are of any importance to meteorology: *rotational motion* (turning of the Earth on its axis), and *revolutional motion* (movement if the Earth in orbit around the Sun).

In the first motion the Earth rotates on its axis once in 24 hours; one-half of the Earth's surface is facing the Sun at all times. The side facing the Sun receives daylight and the side facing away from the Sun is in darkness, accounting for our day and night. The Earth rotates eastward. That's why the Sun rises in the east and sets in the west.

The second motion of the Earth is its revolution around the Sun. This revolution and the tilt of the Earth on its axis are responsible for our seasons. The Earth makes one complete revolution around the Sun in about 365¼ days. The Earth's axis is at an angle of 23½ degrees to its plane of rotation. The Earth's axis points in a nearly fixed direction in space toward the North Star (Polaris) at all times.

Solstices and Equinoxes

When the Earth is in its summer *solstice* (FIG. 1-3), the Northern Hemisphere is inclined at 23½ degrees toward the Sun. This inclination results in more of the Sun's rays reaching the Northern Hemisphere than the Southern Hemisphere. On or about June 21, the Sun shines over the North Pole down the other side to latitude 66½ degrees, the Arctic Circle, and the most vertical rays of the Sun are received at 23½ degrees North latitude, the Tropic of Cancer. The Southern Hemisphere is tilted away from the Sun at this time and the Sun's rays reach only to 66½ degrees South latitude, the Antarctic Circle, and do not go beyond this latitude. The area between the Antarctic Circle and the South Pole is in darkness; the area between the Arctic Circle and the North Pole is receiving the Sun's rays for 24 hours each day.

At the *equinoxes* in March and September, the tilt of the Earth's axis is neither toward nor away from the Sun. For this reason, the Earth receives equal numbers of the Sun's rays in both the Northern and Southern Hemispheres, and the Sun's rays shine most perpendicularly at the Equator.

In December, the situation is exactly reversed from that of June. The South-

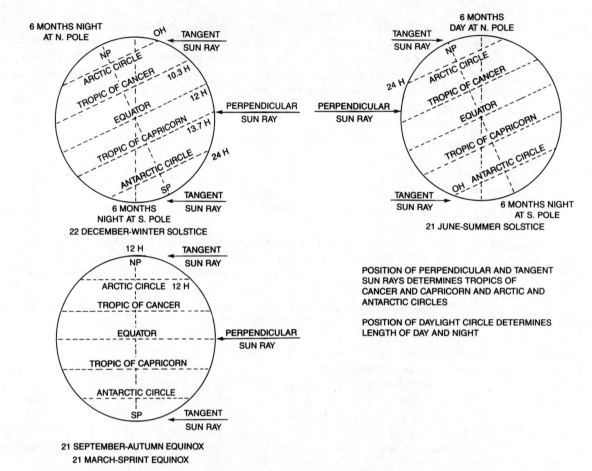

1-3 The Earth's tilt and latitude make the difference in the amount of solar insolation received.

ern Hemisphere now receives more of the Sun's rays. The most perpendicular rays of the Sun are received at 23½ degrees latitude, the Tropic of Capricorn. The south polar area is completely in sunshine and the north polar area is completely in darkness.

Because the revolution of the Earth around the sun is a gradual process, the changes in seasons are gradual. However, it is convenient to mark these changes by specific dates and to identify them by specific names:

March 21, the *vernal equinox*, is when the Earth's axis is vertical to the Sun's rays. Spring begins in the Northern Hemisphere and fall begins in the Southern Hemisphere.

June 21, the *summer solstice*, is when the Earth's axis is inclined 23½ degrees toward the Sun and the Sun has reached its northernmost zenith at the Tropic of Cancer. Summer officially starts in the Northern Hemisphere and winter begins in the Southern Hemisphere.

September 22, the *autumnal equinox*, is when the Earth's axis is again vertical to the Sun's rays. This date marks the beginning of fall in the Northern Hemisphere and spring in the South Hemisphere. It is also the date, along with

March 21, when the Sun reaches its highest position (zenith) directly over the equator.

December 22, the *winter solstice*, is when the Sun has reached its southernmost zenith positions at the Tropic of Capricorn. It marks the beginning of winter in the Northern Hemisphere and the beginning of summer in the Southern Hemisphere.

In some years, the actual dates of the solstices and equinoxes vary by a day from the dates given here because the period of revolution is 365¼ days, and the calendar year is 365 days, except for leap year when it is 366 days.

Zones

Because of its 23½ degrees tilt and its revolution around the Sun, the Earth is marked by five natural light (or heat) zones according to the zone's relative position to the sun's rays. Since the Sun is always at its zenith between the Tropic of Cancer and the Tropic of Capricorn, this is the hottest zone. It is called either the *Equatorial Zone*, the Torrid Zone, the Tropical Zone or simply the Tropics (FIG. 1-4).

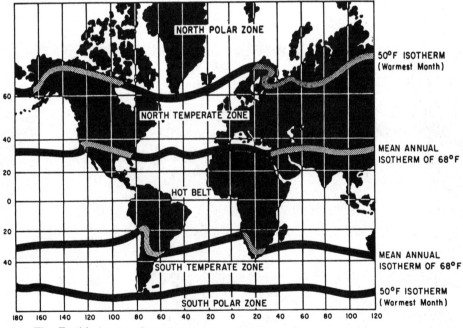

1-4 The Earth's temperature zones.

The zones between the Tropic of Cancer and the Arctic Circle, and between the Tropic of Capricorn and the Antarctic Circle, are the *Temperate Zones*. These zones receive sunshine all year, but less of it in their winters and more in their summers.

The zones between the Arctic Circle and the North Pole and between the Antarctic Circle and the South Pole receive the Sun's rays only for certain parts of the year. At the poles themselves there are six months of darkness and six months of sunshine. This, naturally, makes them the coldest zones. They are, therefore, known as the Frigid or *Polar Zones*.

As you can see, weather is greatly dependent on the relationship of the sun and the Earth. It also depends on solar radiation and how much falls on the Earth.

SOLAR RADIATION

Solar radiation is the total energy given off by the Sun. The Sun's surface beams gamma rays, X-rays, ultraviolet light, visible light, infrared light, heat, and electric waves. Even though the Sun radiates in all wavelengths, about half the radiation is visible light and most of the remainder is infrared. (FIG. 1-5 and 1-6).

Insolation, an acronym for *in*coming *sola*r radia*tion*) is the rate at which solar radiation is received by a surface. There is a wide variety of differences in the amounts of radiation received over the various portions of the earth's surface, as you have just seen. These differences in heating are important in their effect on the weather.

Reflection occurs where a surface turns back a portion of the incident radiation toward its source. About 40 percent of the Sun's radiation bounces off the

1-5 Solar One, the nation's first solar generating plant, was built near Los Angeles, California to produce 10 million watts of energy from the sun's radiation (Dept. of Energy).

1-6 Youngsters are shown how to broil hot dogs in the sunlight using a solar dish lined with aluminum foil. (Dept. of Energy).

Earth's atmosphere and back into space. Earth surfaces also reflect radiation back through the atmosphere, each at different rates. Clouds return about 55 percent, new snow over 80 percent, land from 5 to 30 percent, water 10 percent or less.

The Earth's surface absorbs an average of about 50 percent of the incoming Sun's rays. They warm the Earth and the air through which they pass—the atmosphere—and make the heating and cooling of the Earth uneven. This causes the weather we feel.

LAYERS OF THE ATMOSPHERE

The Earth is surrounded by layers of gases with the densest layer, the troposphere, beginning at the surface and progressively thinning to the exosphere that borders space where there are no gases (FIG. 1-7).

1-7 The Earth's atmosphere.

Troposphere

The *troposphere* is the layer of air around the Earth immediately above its surface. It is about 5½ miles (29,000 feet) thick over the poles, about 7½ miles

(40,000 feet) thick in the mid-latitudes (such as over the United States) and about 11½ miles (61,000 feet) thick over the equator. These are average figures that change somewhat from day to day and from season to season. The troposphere is thicker in summer than in winter and during the day than during the night. All weather, as we know it, occurs in the troposphere.

The troposphere is composed of a mixture of several different gases. By volume, the composition of dry air in the troposphere is 78 percent nitrogen, 21 percent oxygen, nearly 1 percent argon, 0.03 percent carbon dioxide, and traces of helium, hydrogen, neon, krypton, and other elements (FIG. 1-8).

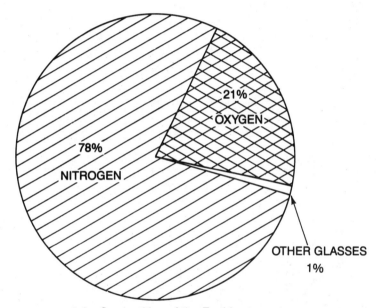

1-8 Composition of the Earth's atmosphere.

The air in the troposphere also contains a variable amount of water vapor. The maximum amount of water vapor that the air can hold depends on the temperature of the air and its pressure. The higher the temperature, the more water vapor it can hold at a given pressure.

The air also contains impurities such as dust, salt particles, soot, and chemicals. These impurities in the air are important because of their effect on visibility and especially because of the part they play in the condensation of water vapor. If the air were absolutely pure, there would be little *condensation* or water vapor turning into water liquid. These small particles act gathering points for the condensation of water vapor.

The temperature in the troposphere usually decreases with height. But there may be inversions for relatively thin layers at any level.

The *tropopause* is a layer between the troposphere and the stratosphere. It is not uniformly thick and is not continuous from the equator to the poles. The tropopause has little or no increase or decrease of temperature with increasing altitude. For most purposes it is considered one with the troposphere.

Stratosphere

The *stratosphere* is directly above the tropopause and extends to about 30 miles (160,000 feet). Temperature varies little with height in the stratosphere through the first 30,000 feet; however, in the upper portion the temperature increases to values nearly equal to surface temperatures. This increase in temperature through the stratosphere is because of the ozone, which absorbs most of the incoming ultraviolet radiation. In fact, in some applications the stratosphere is called the ozonosphere.

The stratopause is the top of the stratosphere. It is the zone marking another reversal of temperature with increasing altitude (FIG. 1-9).

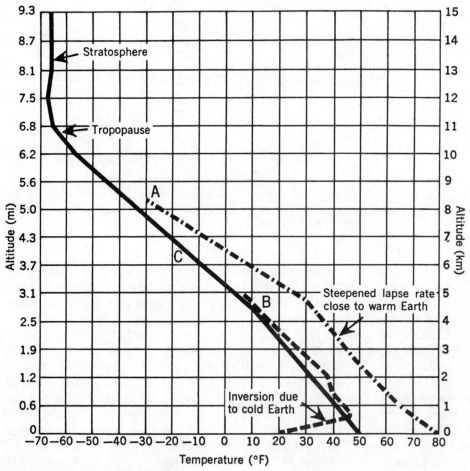

1-9 Variation of temperature with altitude: (A) during a hot sunny day, (B) during a cold clear night, and (C) middle-latitude average.

Mesosphere

The *mesosphere* is a layer about 20 miles (100,000 feet) thick directly over the stratopause. The temperature decreases with height. The mesopause is the thin boundary zone between the mesosphere and the thermosphere. Temperature again increases with altitude. The transitions between layers are very gradual.

Thermosphere

The *thermosphere* is another region in which the temperature increases with height. It extends from the mesopause to outer space. The thermosphere was once called the ionosphere.

Exosphere

The very outer limit of the Earth's atmosphere is called the *exosphere*. It is the zone in which gas atoms are so widely spaced that they rarely collide with one another and have individual orbits around the Earth. There is no weather in the exosphere.

CLIMATE CONTROLS

So why is the weather in New York City different from that in Boise, Idaho? The variation of climatic elements from place to place and from season to season is caused by several factors called climate, or climatic, controls. The same basic factors that cause weather in the atmosphere also determine the climate of an area. These controls act upon temperature, precipitation, humidity, air pressure, and winds to produce many types of weather and climate.

Four factors largely determine the climate of every region on Earth. They are latitude, land and water distribution, topography, and the prevailing elements.

Latitude

Perhaps no other climatic control has such an effect on the weather and climate as does the *latitude*, or the position of a location on the Earth relative to the Sun. The angle at which rays of sunlight reach the earth and the number of "sun" hours each day depend on the distance from the equator (FIG. 1-10).

Regions under direct or nearly direct rays of the Sun receive more heat than those under angled rays. The heat brought about by the slanting rays of early morning may be compared with the heat that is caused by the slanting rays of winter. The heat which is due to the more nearly direct rays of the midday Sun may be compared to the heat resulting from the summer Sun.

The length of the day, like the angle of the Sun's rays, influences the temperature. The length of the day depends on the latitude and the season of the year (FIG. 1-11).

The hot and humid climates of equatorial Africa and South America are good examples of the control that latitude has on climate. At no time during the year are the Sun's rays at much of an angle. Therefore, there is little difference be-

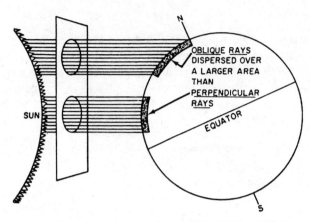

1-10 Different areas of the earth receive different amounts of sunshine.

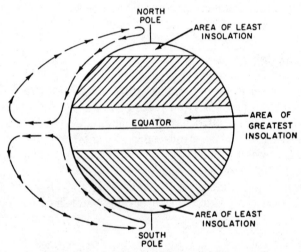

1-11 How air begins to circulate around the earth.

tween the mean temperatures from the coldest and warmest month. Contrast this picture with the opposite extreme in the far north, where the Sun is below the horizon for a great deal of the time during the winter, producing cold temperature that breed powerful polar cold fronts. During the long hours of summer daylight, the Sun's rays make such a small angle with the Earth's surface that the energy received in any giver polar area is extremely small and the sun's effectiveness in minimized. It is, however, enough to thaw lakes and weaken the polar air masses (FIG. 1-12).

Land and Water Distribution

Because land and water heat and cool at different rates, the location of continents and oceans greatly alters the pattern of air temperature and influences the sources and direction of air mass movement.

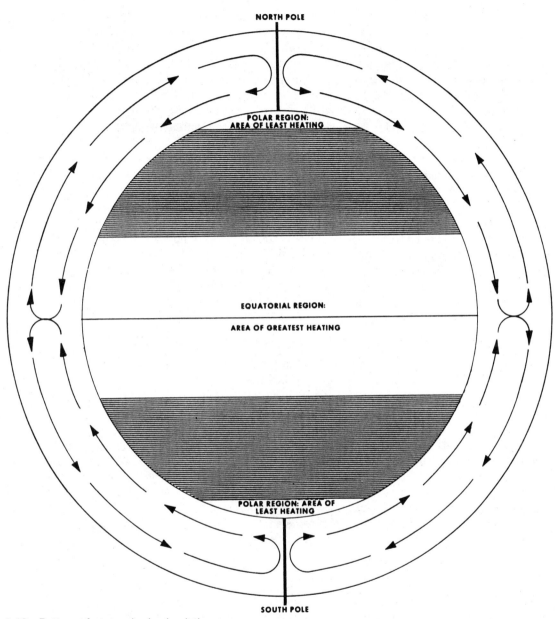

1-12 Pattern of atmospheric circulation.

Coastal areas take on the temperature characteristics of the land or water to their windward. In latitudes of westerly winds, for example, west coasts of continents have oceanic temperatures and east coasts have continental temperatures. The temperatures depend on the windflow.

Since the upper layers of the ocean are nearly always in a state of violent stirring, heat losses or heat gains at the sea surface are distributed throughout a large volume of water. This mixing process sharply reduces the temperature contrasts between day and night, and between winter and summer over oceans.

Land surfaces are not subject to such a mixing process. Thus, violent contrasts between seasons and between day and night are created in the interiors of continents. During winter, a large part of the sparse insolation is reflected back toward space by the snow cover that extends over large portions of the northern continents. For these reasons, the northern area serve as manufacturing plants for dry polar air.

The great temperature difference between land and water surfaces, which reverses between the two seasons, determines to a great extent the seasonal weather patterns.

The nature of the surface affects the local heat distribution, too. Color, texture, and vegetation modify the rate of heating and cooling. Generally, dry surfaces heat and cool faster than moist surfaces. Plowed fields, sandy beaches, and paved roads become hotter than surrounding meadows and wooded areas. During the day, air is warmer over a plowed field than over a forest or swamp. During the night the situation is reversed. The small plane pilot sees this as she fights for control of her plane at low altitudes on warm days.

Topography

Over land, climates may vary radically within a short distance because of land forms and elevations. The height of an area above sea level impacts its climate. For instance, a place located on the equator in the high Andes of South America would have a climate quite different from a place located a few feet above sea level at the same latitude.

Mountains have a powerful influence on climates, especially the long high chains of mountains (such as the Rockies) that act as climatic dividers. These obstacles deflect the tracks of storms and block the passage of air masses on lower levels. If the mass is strong enough to go over a mountain, it will be changed by the time it reaches the other side. East-west mountain chains, such as the Alps, modify south-running polar air and make regions to the south warmer than other points at the same latitude—as is the case in Italy.

Prevailing Elements

Each region of the Earth has its own prevailing elements—air masses, fronts, temperature, wind, and precipitation—that play a part in making weather. These will be covered in the next two chapters to help you understand weather.

DO YOU KNOW...

- The difference between *weather* and *climate*?
- What causes most of our weather?
- The distance between the Earth and the Sun in summer? What makes the Sun so hot?

- The difference between a *solstice* and an *equinox*?
- The names of the layers of the atmosphere? How mountains change the weather?
- What your home's latitude is?

2
Air Masses

THE IDEA OF *AIR MASS* IS ONE OF THE MOST IMPORTANT DEVELOPMENTS IN THE HISTORY OF meteorology and modern forecasting. An *air mass* is a large body of air whose temperature and moisture distribution are the same on all levels. Forecasting is largely a matter of recognizing the air masses, identifying their features, and predicting what they will do.

AIR MASS CLASSIFICATIONS

Air masses have been classified by geography and by temperatures. The geography refers to the source region of the air mass, dividing air masses into four basic categories. These are arctic/antarctic (A), polar (P), tropical (T), and equatorial (E).

Air masses are further divided into *maritime* if it originated over an oceanic surface, or *continental* if the air mass began over a land surface. Maritime arctic/antarctic air masses are rare, since land masses or ice fields affect the polar regions. Most equatorial air masses are maritime.

There are two less common air masses that should be mentioned. The *superior* (S) air mass is generally found over the southwestern United States and the *monsoon* (M) air mass is a local condition in India and southeastern Asia (FIGS. 2-1 and 2-2).

The types of air masses can easily be remembered by using the letters in the word *TAPE* to stand for the first letter of each basic type of air mass: tropical, arctic/antarctic, polar, and equatorial.

Air masses are named for the relative warmth or coldness of the air mass. A

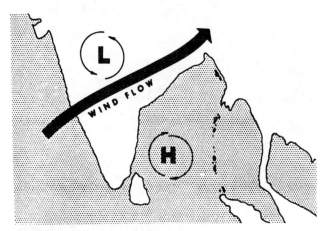

2-1 Pressure system movement causes summer monsoons.

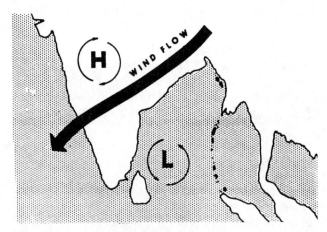

2-2 Winter monsoons.

warm air mass (w) is one that is warmer than the underlying surface. A *cold* air mass (k) is one that is colder than the underlying surface. For example, a continental polar cold air mass is written as *cPk*. An *mTw* is a maritime tropical warm air mass.

Air masses can usually be identified by the type of clouds within them. Cold air masses usually show cumuliform clouds while warm air masses contain stratiform clouds.

Sometimes, and with some air masses, the type may change from night to day. An air mass may show k characteristics during the day and w characteristics at night and vice versa.

SOURCE REGIONS

The air mass *source region* is the area where the air mass originates. The ideal conditions for the birth of an air mass is dormant air over a rather uniform surface (water, land, or icecap) of uniform temperature and humidity.

The sources for the world's air masses are shown in FIG. 2-3. Note that the underlying surfaces are similar. Also note the similar conditions in the various source regions. Compare the southern North Atlantic and Pacific Oceans for maritime tropical air and the deep interiors of North America and Asia for continental polar air.

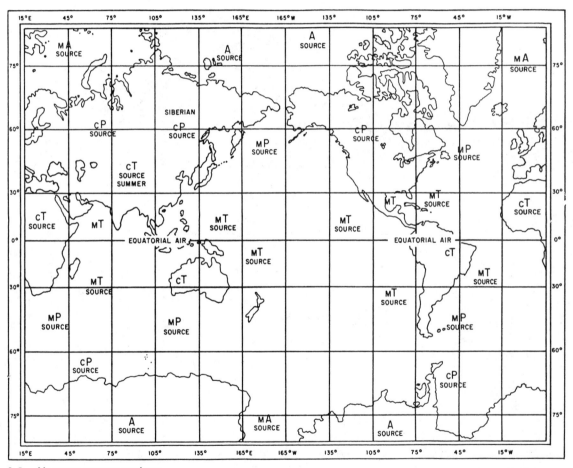

2-3 Air mass source regions.

Arctic air is a permanent high-pressure area near the North Pole, within which is found the arctic air mass source region. In this region there is a gentle flow of air over the polar ice fields, allowing the arctic air mass to form. The air is dry aloft and very cold and stable in the lower altitudes.

Antarctic air is developed in the antarctic region. It is colder at the surface and other levels than arctic air in fall and winter.

Continental polar air source regions include all the land areas affected by the Canadian and Siberian high-pressure cells (FIG. 2-4). In the winter these regions are covered by snow and ice. Because of the intense cold and the absence of water bodies, very little moisture is taken into the air. Note that the word *polar* when applied to air masses does not mean air at the poles. Polar air is generally found between 40 and 60 degrees latitude. Except for ground over northern and central Asia, polar air is generally drier than arctic air.

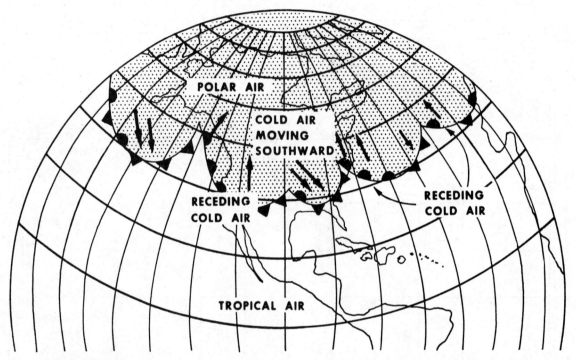

2-4 Polar front.

Maritime polar air source regions include the open unfrozen polar sea areas around 60 degrees latitude, north and south. These areas are sources of moisture for polar air masses. So, air masses forming over these regions are moist, but the moisture is sharply limited by the temperature.

Continental tropical air source regions can be any large land areas lying in the tropical regions, generally between 25 degrees north latitude and 25 degrees south latitude. The large land areas found there are usually desert regions, such as the Sahara and Kalahari Deserts of Africa, the Arabian Desert, and the interior of Australia. The air over these land areas is hot and dry.

Maritime tropical air source regions are the large zones of tropical sea along the belt of high pressure cells that remain in this area most of the year. The air is warm due to low latitude and is able to hold considerable moisture.

Equatorial air's source region is the area from about 10 degrees north to 10 degrees south latitudes surrounding the thermal equator. It is an oceanic belt which is very warm and has a high moisture content. The trade winds from both hemispheres and the isolation over this region causes air to lift to high levels (FIG. 2-5). These conditions cause thunderstorms throughout the year.

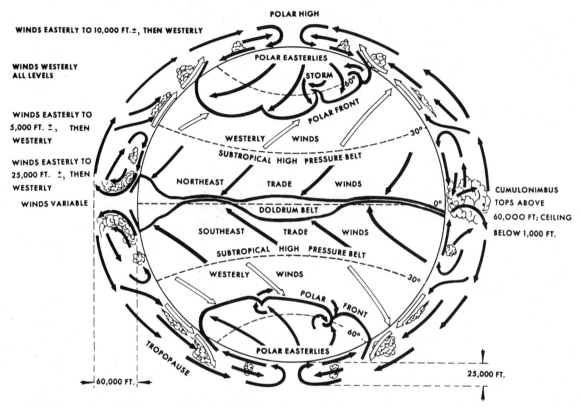

2-5 Idealized pattern of atmospheric circulation.

HOW AIR MASSES CHANGE

When an air mass moves out of its source region, there are a number of factors which act upon it to change it properties (FIG. 2-6). These changes don't occur separately. For instance, when cold air passes over warmer water surfaces, there is not only a release of heat to the air, but also some moisture in the form of rain (FIG. 2-7).

As an air mass expands and slowly moves out of its source region, it travels along a certain path. The surface over which it passes changes the air mass. The way it rotates, clockwise or counterclockwise, also affects how it changes. The time the air mass has been out of its source region also affects the air mass.

Surface Conditions

The first condition that changes an air mass as it leaves its source region is the type and condition of the surface over which it travels. Here, the factors of surface temperature, moisture, and topography are important (FIG. 2-8).

24 AIR MASSES

2-6 Air masses vary in water vapor amounts and dew point.

2-7 When continental polar air of winter moves from cool continent to warm ocean it usually produces precipitation.

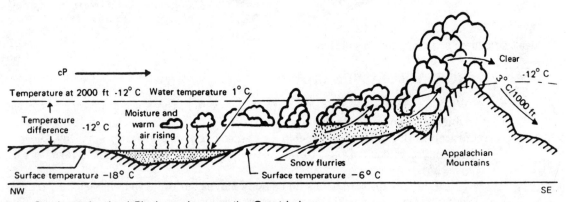

2-8 Continental polar (cP) air moving over the Great Lakes.

The temperature of the surface compared to that of the air mass can change not only the temperature but its balance as well. For example, if the air mass is warm and moves over a colder surface, such as a tropical air moving over colder water, the cold surface cools the lower layers of the air mass and its stability is increased. This balance will extend to the upper layers and fog or low clouds normally occur. If the air mass moves over warmer water, the water heats the lower layers of the air mass, increasing instability. The changes in the air mass show what type of clouds will form as well as the type of precipitation.

The moisture content of the air mass can be changed by evaporation or by condensation and precipitation. If the air mass is moving over continental regions, unfrozen bodies of water can greatly modify the air mass. In the case of an air mass moving from a continent to an ocean, the modification can be considerable. Movement over a water surface will increase the moisture content of the lower layers and the temperature of the surface. For example, the passage of cold air over a warm water surface will result in vertical currents. The passage of warm moist air over a cold surface increases the stability and could result in fog as the air is cooled.

Weather can be affected by mountainous regions. The air mass is changed on the windward side (the side from which the wind blows) by precipitation. As the air descends on the other side of the mountain the air becomes warmer, drier and more stable (FIG. 2-9).

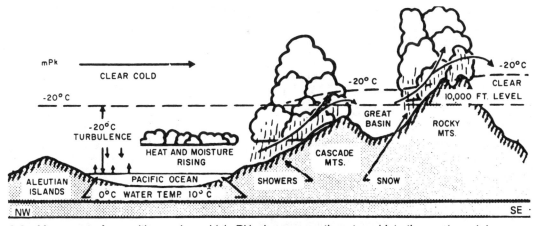

2-9 Movement of a maritime polar cold (mPk) air mass southeastward into the western states.

Path

After an air mass has left its source region, the path it follows and its turning clockwise or counterclockwise have a great effect on its stability. If the air mass circles counterclockwise in the Northern Hemisphere (clockwise in the Southern Hemisphere), its stability in the upper levels is decreased. On the other hand, if it circles clockwise in the Northern Hemisphere (counterclockwise in the Southern Hemisphere), it is more stable in the upper levels.

Age

Although the age of an air mass cannot change the air, it will determine how much change will take place. For example, an air mass that has recently moved from its source region will not have had time to change very much. However, an air mass that has moved into a new region and remained motionless for some time, it will soon lose many of its original qualities.

Table 2-1 shows the modifying process, what takes place, and the change in the air mass. Remember, these processes do not occur independently but two or more processes usually occur at the same time. Remember that conditions shown are only average conditions and that cases may be quite different. However, understanding the types, sources, and changing factors in air masses will give you a greater understanding of weather and how to forecast it accurately.

Table 2-1 Thermal and Mechanical Air Mass Modifications

The Process	What takes place	Results
A. THERMAL		
1. Heating from below.	Air mass passes from over a cold surface to a warm surface, or surface under air mass is heated by sun.	Decrease in stability.
2. Cooling from below.	Air mass passes from over a warm surface to a cold surface, OR radiational cooling of surface under air mass takes place.	Increase in stability.
3. Addition of moisture.	By evaporation from water, ice, or snow surfaces, or moist ground, or from raindrops or other precipitation which falls from overrunning saturated air currents.	Decrease in stability.
4. Removal of moisture.	By condensation and precipitation from the air mass.	Increase in stability.
B. MECHANICAL		
1. Tuburlent mixing	Up- and down-draft.	Tends to result in a thorough mixing of the air through the layer where the turbulence exists.
2. Sinking.	Movement down from above colder air masses or descent from high elevations to lowlands, subsidence and lateral spreading.	Increases stability.
3. Lifting.	Movement up over colder air masses or over elevations of land or to compensate for air at the same level converging.	Decreases stability.

UNDERSTANDING FRONTS

In weather, the term *front* means a boundary separating two different air masses. From this definition, you can see the close relationship between air masses and fronts. In fact, without the air masses, there would be no fronts.

The centers of action bring together air masses of different physical properties. The region of change between two air masses is called a *frontal zone*. The main frontal zones of the Northern Hemisphere are the arctic frontal zone and the polar frontal zone. The most important frontal zone affecting the United

States is the polar front. The polar front occurs between the cold polar air and warm tropical air. During the winter months, the polar front pushes farther southward (due to the greater density of the polar air) than during the summer months. In the summer months, the front seldom moves farther south than the central United States.

On a surface map, a front is shown by a line separating two air masses. This is only a picture of the surface conditions. Figure 2-10 is a vertical look at a front.

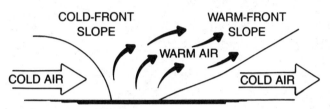

2-10 Vertical view of a frontal system (without clouds shown).

A cold air mass, being heavier, tries to push under a warm air mass. Thus, the cold air is below and the warm air is above the surface.

The dew point is normally more consistent than the temperature through the day, except with the passage of a front. Therefore, the dew point is a good indicator that a front has passed.

COLD FRONTS

A *cold front* is the line where a wedge of cold air is under-running and displacing a warmer air mass. There are certain weather conditions that are typical of cold fronts. In general, the temperature and humidity decrease, the pressure rises and, in the Northern Hemisphere, the wind shifts (usually from southwest to northwest) with the passage of a cold front. How dense rain and snow fall depends on the vertical speeds in the warm air mass. Therefore, cold fronts are either slow-moving or fast-moving cold fronts (FIG. 2-11).

Slow-Moving Cold Fronts

With the slow-moving front warm air generally moves upward along the entire frontal surface except for lifting along the lower portion of the front. The cloud and precipitation area is large. It can be recognized by cumulonimbus and nimbostratus clouds, with showers and thunderstorms at and immediately to the rear of the surface front. This area is followed by a region of rain and nimbostratus clouds merging into a region of altostratus clouds. Then come the cirrostratus clouds, which can extend several hundred miles behind the surface front.

The development of clouds, showers, and thunderstorms depends on the conditions within the warm air mass. Within the cold air mass there might be some stratified clouds in the rain area, but there are no clouds beyond this area

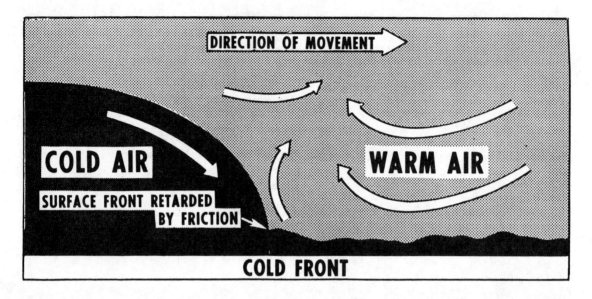

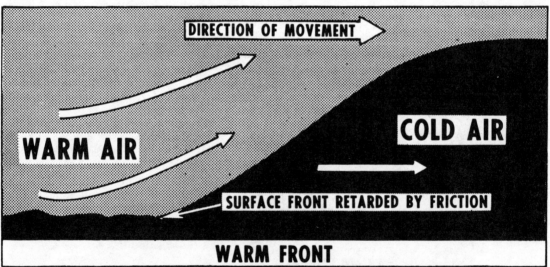

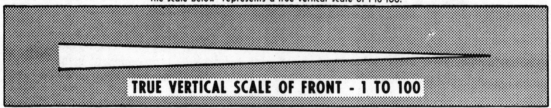

2-11 Vertical cross section of cold and warm fronts showing frontal slopes.

unless it is unstable. In the latter case, some cumulus clouds might develop. This type of front is slow moving; 15 knots (17 miles an hour) is average. Figure 2-12 shows a cross section of a typical slow-moving cold front.

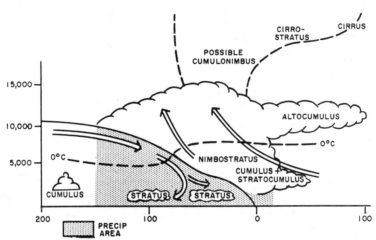

2-12   Vertical cross section of a slow-moving cold front.

Fast-Moving Cold Fronts

With the fast-moving cold front the high level warm air along the frontal surface descends and the warm air near the surface is pushed vigorously upward. This type of front usually moves rapidly; 25 to 30 knots (29 to 34 miles an hour) is the average speed. Because of this, there is a relatively narrow but often violent band of weather with the passing of a fast-moving cold front. If the warm air mass is unstable and moist, cumulonimbus clouds, showers, and thunderstorms occur just ahead of the surface front. Rapid clearing occurs behind the front. Frequently, altostratus and altocumulus cloud layers form and drift ahead of the main cloud band. The more unstable the warm air mass, the more violent the weather. If the warm air is relatively dry, this type of front might not cause precipitation or clouds. It is with the fast-moving cold front that squall lines (violent winds and precipitation) are associated.

Figure 2-13 shows a typical cross section of a fast-moving cold front. It also shows the cloud shield, precipitation shield and frontal slope typically found this type of front.

WARM FRONTS

A *warm front* is the line between the forward edge of an advancing mass of relatively warm air and a retreating, colder air mass. As in the case of the cold front, this term is incorrectly used to refer to a warm frontal surface.

As a warm front passes, the winds shift from southeast to southwest or west, but the shift is not as pronounced as with the cold front. The temperatures are colder ahead of the front. The cloud formation is somewhat unusual: cirrus,

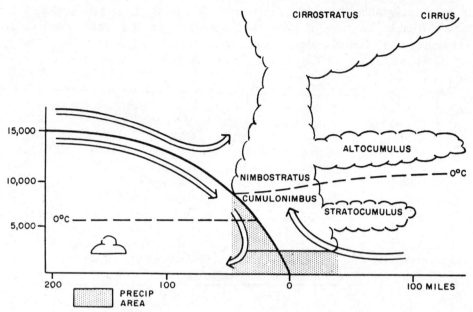

2-13 Vertical cross section of a fast-moving cold front.

cirrostratus, altostratus, nimbostratus, and stratus. The cirrus clouds might appear 700 to 1000 miles ahead of the surface front.

Rain or light showers, snow, or drizzle is frequent as much as 300 miles in advance of the surface front. Surface precipitation is from nimbostratus clouds in the warm air above the frontal surface and from stratus clouds in the cold air. However, when the warm air is unstable, showers and thunderstorms may occur in addition to the steady precipitation. Fog is common in the cold air ahead of a warm front.

Clearing usually occurs after the passage of a warm front, but under some conditions drizzle and fog can occur within the warm sector. Warm fronts usually move in the direction of the isobars of the warm sector; in the Northern Hemisphere this is usually east to northeast. Their speed of movement is normally less than that of cold fronts. On the average it is about 10 knots (12 miles an hour). Figure 2-14 shows the main features of warm fronts under average conditions.

OCCLUDED FRONTS

An *occluded front* appears when a cold front overtakes a warm front. One of the two fronts is lifted aloft and the warm air between the fronts is shut off from the Earth's surface. An occluded front is often referred to as an *occlusion*, a closing. The type of occlusion depends on the temperature difference between the cold air in advance of the warm front and the cold air behind the cold front.

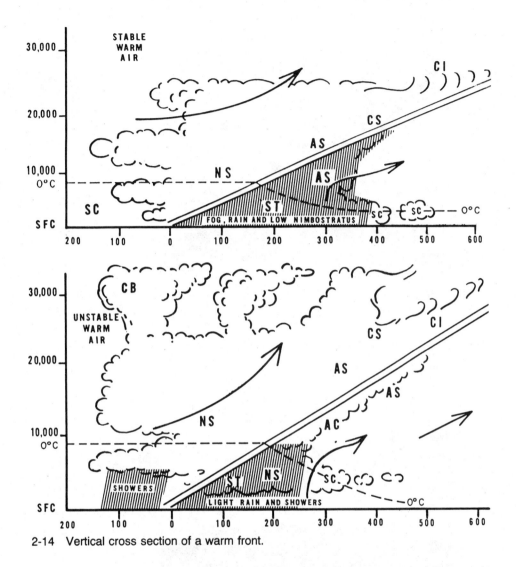

2-14 Vertical cross section of a warm front.

Figure 2-15 shows a warm-type occlusion. If the air in advance of the warm front is colder than the air behind the cold front, the cold front rides up the warm frontal slope. If the cold air ahead of the warm front is warmer than the cold air behind the cold front, the cold frontal surface underruns the warm front and the occluded front is called a cold-type occlusion (FIG. 2-16).

The difference between a warm-type and cold-type occlusion is the location of the upper front relative to the surface front (FIG. 2-17). In a warm-type occlusion, the upper cold front comes before the surface front by as much as 200 miles. In the cold-type occlusion, the upper warm front follows the surface occluded front by 20 to 50 miles.

Since the occluded front is a combination of a cold front and a warm front, the weather is the worst of both. The cold front's narrow band of violent weather

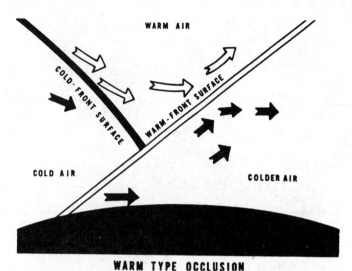

2-15 Vertical cross section of a warm type occlusion.

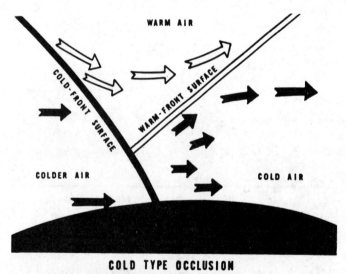

2-16 Vertical cross section of a cold type occlusion.

and the warm front's wide-spread area of cloudiness and precipitation all occur in combination along the occluded front. The moist violent weather occurs at the tip of the occlusion, the point at which the cold front is overtaking the warm front.

STATIONARY FRONTS

When a front is *stationary*, the cold air mass, as a whole, does not move either toward or away from the front. In terms of wind direction this means that the wind blows neither toward not away from the front, but parallel to it. It follows

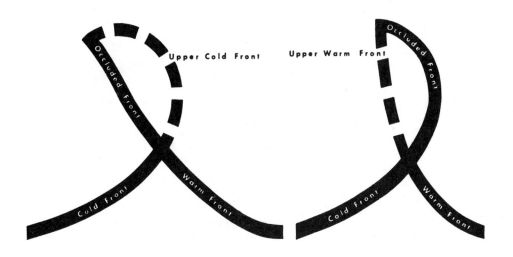

(A) WARM TYPE OCCLUSION **(B) COLD TYPE OCCLUSION**

2-17 Occlusions (in the horizontal) and associated upper fronts.

that the isobars, lines of identical barometric pressure, are also nearly parallel to a stationary front. This makes it easy to recognize a stationary front on a weather map.

The frictional inflow of warm air toward a stationary front causes a slow upglide of air on the frontal surface. As the air is lifted to and beyond its condensation level, clouds form in the warm air above the front.

If the warm air in a stationary front is stable, the clouds are stratiform. Drizzle might fall. As the air is lifted beyond the freezing level, icing conditions develop and light rain or snow might fall. At very high levels above the front ice clouds are present (FIG. 2-18).

If the warm air is unstable and enough lifting occurs, the clouds are then cumuliform, or stratiform with cumuliform sections. If the energy release is

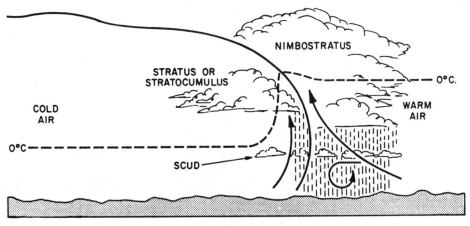

2-18 Vertical cross section of a steep stable stationary front.

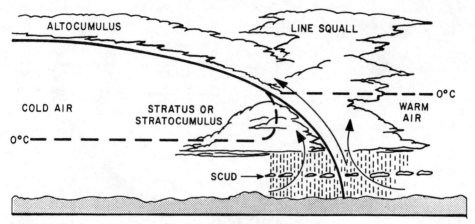

2-19   Vertical cross section of a steep unstable stationary front.

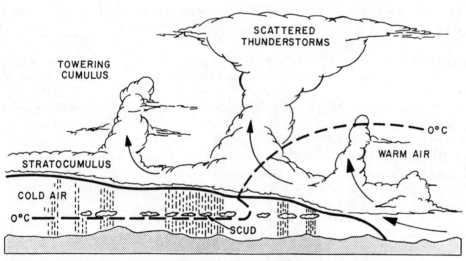

2-20   Vertical cross section of a shallow unstable stationary front.

great (warm, moist, unstable air) thunderstorms result. Rainfall is generally showery (FIGS. 2-19 and 2-20).

Within the cold air mass, extensive fog and low ceiling might result. The cold air is saturated by warm rain or drizzle falling through it from the warm air mass above. If the temperature is below 0°C (32°F), icing might occur, but generally it is light (FIG. 2-21).

The width of the band of precipitation and low ceiling varies from 50 miles to about 200 miles, depending on the slope of the front and the temperatures of the air masses. Unfortunately for travelers, a stationary front can slow airport operations by staying in the area for several days (FIG. 2-22).

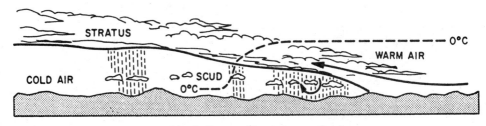

2-21 Vertical cross section of a shallow stable stationary front.

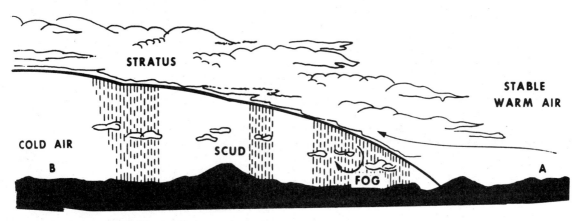

2-22 Stationary front with stable warm air.

LOWS AND HIGHS

Moving air masses of warmer and colder air, along with their fronts, cause changes in the barometric pressure. In fact, fronts can be found by drawing a connecting line on a weather map between points of low barometric pressure—called a *trough*.

A *low* is a pressure system in which the barometric pressure decreases toward the center and the windflow around the system is counterclockwise in the Northern Hemisphere (FIG. 2-23). The terms low, and cyclone or wave cyclone mean the same thing (FIG. 2-24). Any pressure system in the Northern Hemisphere with a counterclockwise (cyclonic) windflow is a *cyclone*. The term "cyclone" is inaccurately used to describe some violent storms. Low-pressure systems with severe storm characteristics are more accurately called hurricanes, typhoons, tropical storms, tornadoes, or waterspouts to identify the exact nature of the storm.

Low-pressure systems usually bring cloudy weather with rain or snow and move through quickly. Some lows, called tropical lows, bring warm weather. Thermal lows are caused by intense surface heating and resulting low air density over barren continental areas. They are relatively dry with few clouds and practically no precipitation. These thermal lows are almost stationary and predominate over continental areas in the summer.

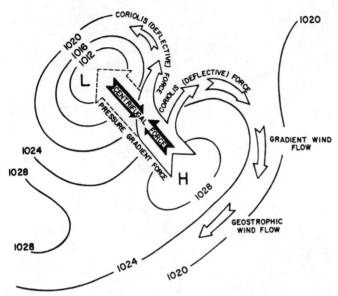

2-23 Examples of circulation about high and low pressure systems.

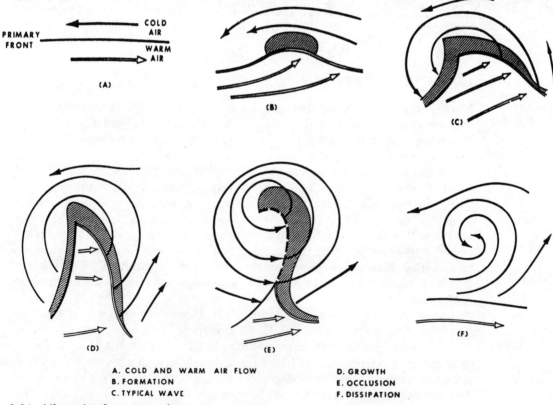

A. COLD AND WARM AIR FLOW
B. FORMATION
C. TYPICAL WAVE
D. GROWTH
E. OCCLUSION
F. DISSIPATION

2-24 Life cycle of a wave cyclone.

A high-pressure system in which the barometric pressure increases toward the center and the windflow around the system is clockwise in the Northern Hemisphere (FIG. 2-25). The terms *high* and *anticyclone* (opposite of cyclone) mean the same thing. Any pressure system in the Northern Hemisphere with a clockwise (anticyclonic) windflow is an anticyclone or high (FIG. 2-26).

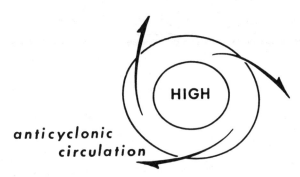

2-25   Anticyclone (high-pressure) system circulation.

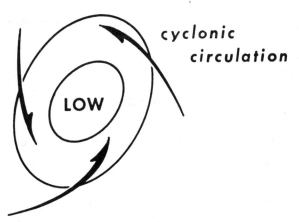

2-26   Cyclone (low-pressure) system circulation.

High-pressure systems are often found over cold surfaces where the air is dense. They are more intense over continental areas in winter and oceanic areas in summer. They usually bring fair weather and stay awhile, offering clearer skies, less winds, and better visibility (FIG. 2-27).

You will also hear three other terms when discussing highs and lows: ridge, trough, and col. A *col* is a saddleback region between two highs or two lows. The weather is erratic and unpredictable. A *trough* is a long area of low pressure with the lowest pressure along the trough line, where weather is frequently violent. A *ridge* is an area of high pressure with highest pressure along the ridge line. It usually offers good weather.

Air masses and fronts both make and reflect changes in the elements of the

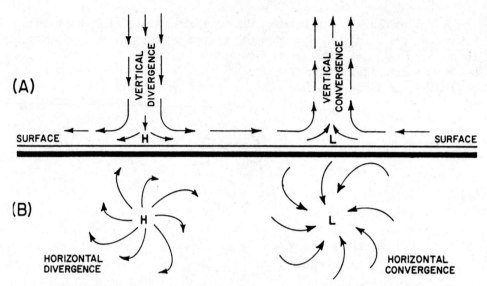

2-27 Horizontal and vertical views at the movement of high and low-pressure systems in the Northern Hemisphere.

atmosphere. Learning more about these elements will improve your understanding of the weather.

DO YOU KNOW...

- The two ways that air masses are classified?
- The source regions for weather in your area?
- The difference between a cold air mass and a warm air mass?
- The meaning of *occluded*?
- How to tell a *low* from a *high*?
- Why learning about air masses is important?

3
The Elements

WHAT WE CALL *THE WEATHER* CAN BE BROKEN DOWN INTO SIX ELEMENTS: TEMPERATURE, pressure, wind, moisture, clouds, and precipitation. These are measurable forces in the earth's atmosphere that can be used to read and forecast weather. These elements combine in many ways to give us clear and temperate spring days, humid nights, Christmas snowstorms, and red sunsets.

TEMPERATURE

All substances are made up of super-small molecules, which are in more or less rapid motion. Temperature is actually a measure of the average speed of these molecules. As the speed of these molecules increases in a substance under constant pressure, the temperature of the substance increases. Surface air temperatures are usually measured with liquid-in-glass thermometers (to be discussed in Chapters 5 on basic instruments).

Two fixed temperatures, the melting point of ice and the boiling point of water (at standard atmospheric pressure), are used to calibrate thermometers. The two scales in common use are *centigrade* (or Celsius) and *Fahrenheit* (FIG. 3-1). The centigrade scale was devised by Anders Celsius during the 18th century. The terms centigrade scale and degrees centigrade have been used for many years, but the accepted terminology now is *Celsius* scale and degrees Celsius (°C).

To convert from Celsius to Fahrenheit, use the equation:

$$F = 9/5(°C + 32)$$

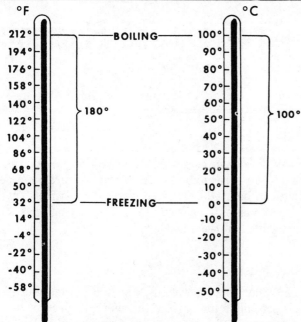

3-1 Celsius and Fahrenheit temperature scale and conversion table.

To convert from Fahrenheit to Celsius, use the equation:
$$C = 5/9\,(°F - 32)$$

Aviators know that there is normally an overall decrease of temperature in the troposphere as an aircraft gains altitude, because the air nearest the ground (which has been heated by incoming solar radiation) receives the largest amount of heat. The change in temperature with altitude is called the *temperature lapse rate* and is usually expressed in degrees per thousand feet. On one day, the air may have a decrease in temperature of 3°C for each thousand feet gained in altitude. Another day may show a decrease of 1°C per thousand feet. A third day may reveal the temperature increasing for a distance of one or two thousand feet above the ground, called and inverted lapse rate or inversion, and then decreasing at the rate of 3°C per thousand feet. If such observations taken day after day over thousands of locations on the earth were averaged, the average lapse rate would be about 2°C decrease per thousand feet.

The change in the lapse rate is the main reason that temperatures aloft, or high in the atmosphere, are normally measured twice daily. Valuable information can be gathered from observations aloft such as freezing level, types of clouds that will form, maximum and minimum surface temperatures for the day, and the amount of air turbulence. The rate of change varies from day to day depending in the amount of heat reaching the Earth, the amount escaping from the Earth, and the type and amount of advection (warmer or colder air moving into the area from other regions).

CAUSES OF TEMPERATURE CHANGE

Three important factors are responsible for temperature changes on the earth's surface. They are the Earth's daily rotation about its axis, yearly revolution around the sun, and variations in land mass and water mass heating (FIGS. 3-2 through 3-4).

Daily surface heating and cooling are the result of the Earth's rotation about its axis. As the Earth turns, the side facing the Sun is heated and the side away

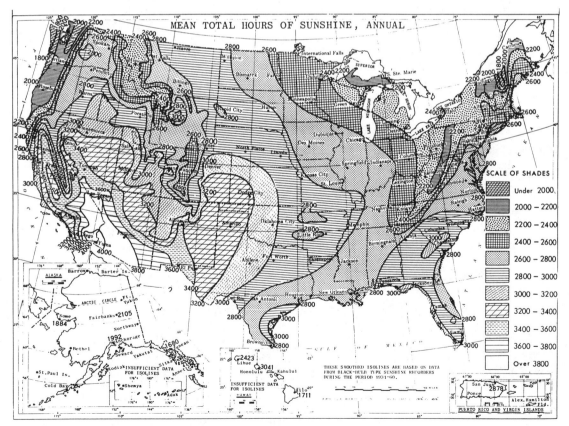

3-2 Mean annual total hours of sunshine.

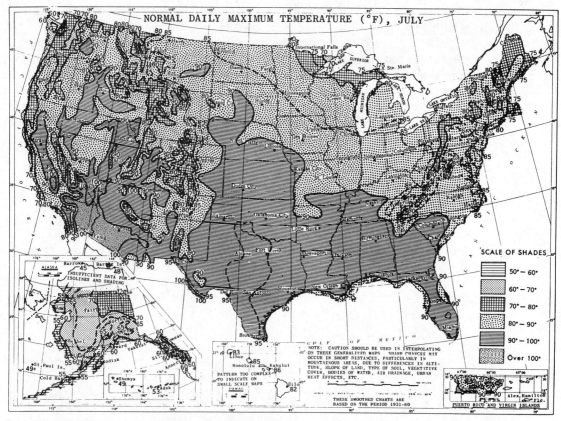

3-3 Normal daily maximum temperature (°F) for July.

from the Sun is cooled. Generally, the lowest temperature occurs near sunrise and the highest temperature is recorded between 1:00 and 4:00 p.m.

The affects of the yearly revolution around the sun are modified by the tilt of the axis of the Earth. Areas under the direct rays of the sun receive more heat than those under rays coming in at an angle. One factor affecting this is shown in FIG. 3-5. The lines intersecting the ground represent a group of rays from the Sun, which in turn represents a given amount of energy. As these rays strike the ground, as in winter, the energy is distributed over an area whose width is from A to C. In summer, when the Sun is closer to the overhead position, the same group of rays fall on an area whose width is from B to C, which is much less than A to C. Thus, the Sun's energy is more concentrated in summer than in winter, which results in greater heating per unit area.

Another reason for the lesser heating in winter is that the angled rays pass through more atmosphere. This absorbs, reflects, and scatters the Sun's energy. Fewer of these angled rays reach the Earth's surface or lower layers of the atmosphere.

The third factor is variation in land mass and water mass heating. Land areas heat and cool more rapidly than water areas; water tends to have a more

3-4 Normal daily minimum temperature (°F) for January.

uniform temperature throughout the year. During the night, water retains its warmth while the land mass rapidly loses its heat to space. This difference between land and water heating rates also influences seasonal temperatures. That means oceanic climates are warmer in winter and cooler in summer than land climates at the same latitude.

Another factor in considering temperature and its role in weather forecasting is heat transfer. When a solid object or volume of liquid or gas loses more heat energy than it gains its temperature decreases. When it acquires more heat energy than it loses its temperature increases. At times, a complete change in the character of the weather occurs over land areas between early morning and mid-afternoon. This change takes place because of the heating and cooling of the earth and its atmosphere. The four effective methods of heat transfer in the atmosphere are radiation, conduction, convection, and advection.

Radiation is the process that transfers energy through space or a material from one location to another. The Sun radiated heat energy to the Earth, called insolation. Any reflected energy is called terrestrial radiation. The *greenhouse effect* is the ability of water vapor, smoke, haze, and clouds to reduce the cooling of the Earth during the night.

44 THE ELEMENTS

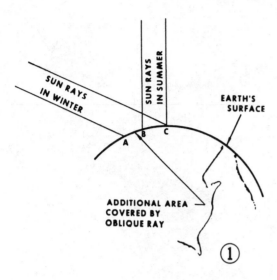

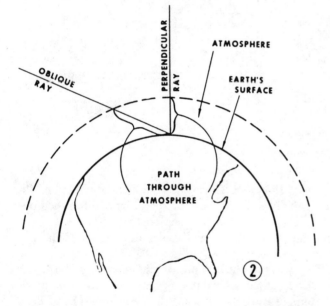

3-5 Variations in solar energy received by the earth.

Conduction is the transfer of heat by contact. This process is important in meteorology because it causes the air close to the surface of the Earth to heat during the day and cool during the night.

Convection is the heat transferred when air moves upward or downward in the atmosphere, either because of solar heating (thermal convection) or air being forced to go over a mountain range (mechanical convection).

Advection is the horizontal movement of air (wind) that causes heating or cooling of an area. Understanding the different factors that change the temper-

Table 3-1 Highest Temperatures in the U.S. by Month.

Month	Temp. (°F)	Year	Day	State	Place	Elevation (feet)
Jan.	98	1936	17	Texas	Laredo	421
Feb.*	105	1963	3	Ariz.	Montezuma	735
Mar.*	108	1954	31	Texas	Rio Grande City	168
Apr.	118	1898	25	Calif.	Volcano Springs	−220
May*	124	1896	27	Calif.	Salton	−263
June*†	127	1896	15	Ariz.	Ft. Mohave	555
July	134	1913	10	Calif.	Greenland Ranch	−178
Aug.†	127	1933	12	Calif.	Greenland Ranch	−178
Sept.	126	1950	2	Calif.	Mecca	−175
Oct.*	116	1917	5	Ariz.	Sentinel	685
Nov.*	105	1906	12	Calif.	Croftonville	1759
Dec.	100	1938	8	Calif.	La Mesa	538

*Two or more occurrences, most recent given.
†Slightly higher temperatures in old records are not used because of lack of information on exposure of instruments.

Table 3-2 Lowest Temperatures in the U.S. by Month.

Month	Temp. (°F)	Year	Day	State	Place	Elevation (feet)
Jan.†	−70	1954	20	Mont.	Rogers Pass	5,470
Feb.	−66	1933	9	Mont.	Riverside R. S.	6,700
Mar.	−50	1906	17	Wyo.	Snake River	6,862
Apr.	−36	1945	5	N. Mex.	Eagle Nest	8,250
May	−15	1964	7	Calif.	White Mountain 2	12,470
June	2	1907	13	Calif.	Tamarack	8,000
July*	10	1911	21	Wyo.	Painter	6,800
Aug.*	5	1910	25	Mont.	Bowen	6,080
Sept*	−9	1926	24	Mont.	Riverside R. S.	6,700
Oct.	−33	1917	29	Wyo.	Soda Butte	6,600
Nov.	−53	1959	16	Mont.	Lincoln 14 NE	5,130
Dec.*	−59	1924	19	Mont.	Riverside R. S.	6,700

† −80°F, Jan. 23, 1971 at Prospect Creek Camp, Alaska, elevation 1,100 ft.
*Two or more occurrences, most recent given.

ature can help you in understanding weather elements. (For record high and low temperatures, see TABLES 3-1 through 3-4).

PRESSURE

No matter where you stand on this Earth, there's a certain amount of atmospheric pressure holding you, and everything else, down. Technically, atmospheric pressure is the pressure exerted by the atmosphere as a result of gravitational attraction acting on the column of air lying directly above a point. Because of the constant and complex air movements and changes in temperature and moisture content of the air, the weight of the air column over a fixed point is

Table 3-3 Highest Recorded Temperatures by State.

State	Temp. (°F)	Date	Station	Elevation (feet)
Ala.	112	Sept. 5, 1925	Centerville	345
Alaska	100	June 27, 1915	Fort Yukon	—
Ariz.	127	July 7, 1905*	Parker	345
Ark.	120	Aug. 10, 1936	Ozark	396
Calif.	134	July 10, 1913	Greenland Ranch	−178
Colo.	118	July 11, 1888	Bennett	—
Conn.	105	July 22, 1926	Waterbury	400
Del.	110	July 21, 1930	Millsboro	20
D.C.	106	July 20, 1930*	Washington	112
Fla.	109	June 29, 1931	Monticello	207
Ga.	112	July 24, 1952	Louisville	337
Hawaii	100	Apr. 27, 1931	Pahala	850
Idaho	118	July 28, 1934	Orofino	1,027
Ill.	117	July 14, 1945	E. St. Louis	410
Ind.	116	July 14, 1936	Collegeville	672
Iowa	118	July 20, 1934	Keokuk	614
Kans.	121	July 24, 1936*	Alton (near)	1,651
Ky.	114	July 28, 1930	Greensburg	581
La.	114	Aug. 10, 1936	Plain Dealing	268
Maine	105	July 10, 1911*	North Bridgton	450
Md.	109	July 10, 1936*	Cumberland & Frederick	623;325
Mass.	106	July 4, 1911*	Lawrence	51
Mich.	112	July 13, 1936	Mio	963
Minn.	114	July 6, 1936*	Moorhead	904
Miss.	115	July 29, 1930	Holly Springs	600
Mo.	118	July 14, 1954*	Warsaw & Union	687;560
Mont.	117	July 5, 1937	Medicine Lake	1,950
Nebr.	118	July 24, 1936*	Minden	2,169
Nev.	122	June 23, 1945*	Overton	1,240
N.H.	106	July 4, 1911	Nashua	125
N.J.	110	July 10, 1936	Runyon	18
N. Mex.	116	July 14, 1934*	Orogrande	4,171
N.Y.	108	July 22, 1926	Troy	35
N.C.	109	Sept. 7, 1954*	Weldon	81
N.Dak.	121	July 6, 1936	Steele	1,857
Ohio	113	July 21, 1934*	Gallipolis (near)	673
Okla.	120	July 26, 1943*	Tishomingo	670
Oreg.	119	Aug. 10, 1898*	Pendleton	1,074
Pa.	111	July 10, 1936*	Phoenixville	100
R.I.	102	July 30, 1949	Greenville	420
S.C.	111	June 28, 1954*	Camden	170
S.Dak.	120	July 5, 1936	Gannvalley	1,750
Tenn.	113	Aug. 9, 1930*	Perryville	377
Tex.	120	Aug. 12, 1936	Seymour	1,291
Utah	116	June 28, 1892	Saint George	2,880
Vt.	105	July 4, 1911	Vernon	310
Va.	110	July 15, 1954	Balcony Falls	725
Wash.	118	Aug. 5, 1961*	Ice Harbor Dam	475
W.Va.	112	July 10, 1936*	Martinsburg	435
Wis.	114	July 13, 1936	Wisconsin Dells	900
Wyo.	114	July 12, 1900	Basin	3,500
P.R.	103	Aug. 22, 1906	San Lorenzo	203

*Also on earlier dates at the same or other places in the state.

Table 3-4 Lowest Recorded Temperatures by State.

State	Temp. (°F)	Date	Station	Elevation (feet)
Ala.	−24	Jan. 31, 1966	Russellville	880
Alaska	−80	Jan. 23, 1971	Prospect Creek Camp	1,100
Ariz.	−40	Jan. 7, 1971	Hawley Lake	8,180
Ark.	−29	Feb 13, 1905	Pond	1,250
Calif.	−45	Jan. 20, 1937	Boca	5,532
Colo.	−60	Feb. 1, 1951	Taylor Park	9,206
Conn.	−32	Feb. 16, 1943	Falls Village	585
Del.	−17	Jan. 17, 1893	Millsboro	20
D.C.	−15	Feb. 11, 1899	Washington	112
Fla.	−2	Feb. 13, 1899	Tallahassee	193
Ga.	−17	Jan. 27, 1940	CCC Camp F-16	−
Hawaii	14	Jan. 2, 1961	Haleakala, Maui Island	9,750
Idaho	−60	Jan. 18, 1943	Island Park Dam	6,285
Ill.	−35	Jan. 22, 1930	Mount Carroll	817
Ind.	−35	Feb. 2, 1951	Greensburg	954
Iowa	−47	Jan. 12, 1912	Washta	1,157
Kans.	−40	Feb. 13, 1905	Lebanon	1,812
Ky.	−34	Jan. 28, 1963	Cynthiana	684
La.	−16	Feb. 13, 1899	Minden	194
Maine	−48	Jan. 19, 1925	Van Buren	510
Md.	−40	Jan. 13, 1912	Oakland	2,461
Mass.	−34	Jan. 18. 1957	Birch Hill Dam	840
Mich.	−51	Feb. 9, 1934	Vanderbilt	785
Minn.	−59	Feb. 16, 1903*	Pokegama Dam	1,280
Miss.	−19	Jan. 30, 1966	Corinth 4 SW	420
Mo.	−40	Feb. 13, 1905	Warsaw	700
Mont.	−70	Jan. 20, 1954	Rogers Pass	5,470
Nebr.	−47	Feb. 12, 1899	Camp Clarke	3,700
Nev.	−50	Jan. 8, 1937	San Jacinto	5,200
N.H.	−46	Jan. 28, 1925	Pittsburg	1,575
N.J.	−34	Jan. 5, 1904	River Vale	70
N.Mex.	−50	Feb. 1, 1951	Gavilan	7,350
N.Y.	−52	Feb. 9, 1934	Stillwater Reservoir	1,670
N.C.	−29	Jan. 30, 1966	Mt. Mitchell	6,525
N.Da.	−60	Feb. 15, 1936	Parshall	1,929
Ohio	−39	Feb. 10, 1899	Milligan	800
Okla.	−27	Jan. 18, 1930*	Watts	958
Oreg.	−54	Feb. 10, 1933*	Seneca	4,700
Pa.	−42	Jan. 5, 1904	Smethport	−
R.I.	−23	Jan. 11, 1942	Kingston	100
S.C.	−13	Jan. 26, 1940	Longcreek (near)	1,631
S.Dak.	−58	Feb. 17, 1936	McIntosh	2,277
Tenn.	−32	Dec. 30, 1917	Mountain City	2,471
Tex.	−23	Feb. 8, 1933*	Seminole	3,275
Utah	−50	Jan. 5, 1913*	Strawberry Tunnel	7,650
Vt.	−50	Dec. 30, 1933	Bloomfield	915
Va.	−29	Feb. 10, 1899	Monterey	−
Wash.	−48	Dec. 30, 1968	Mazama & Winthrop	2,120;1,755
W.Va.	−37	Dec. 30, 1917	Lewisburg	2,200
Wis.	−54	Jan. 24, 1922	Danbury	908
Wyo.	−63	Feb. 9, 1933	Moran	6,700
P.R.	40	Mar. 9, 1911	Aibonito	2,059

*Also on earlier dates at the same or other places in the state.

always changing. These changes in weight (and therefore pressure) are unfelt by us, but are measurable with pressure-sensitive instruments. The instruments commonly used in measuring atmospheric pressure are the mercurial barometer and the aneroid barometer.

Inches of mercury and *millibars* are two pressure units commonly used. The most common unit is the *inch of mercury* (abbreviated Hg) derived from the height of the mercury column in a mercury barometer. In meteorology, it is more convenient to express the pressure directly as a force per unit area, the *millibar* (abbreviated mb) which is atmospheric pressure equal to a force of 1000 dynes per square centimeter. Standard atmospheric pressure at sea level is 1013.2 millibars, which is equivalent to 29.92 inches of mercury (FIG. 3-6). The pressure exerted by one inch of mercury is about 34 millibars, and the pressure exerted by one millibar is about 0.03 inch of mercury.

To eliminate pressure variations caused by stations being at different altitudes, the mean sea level pressure is plotted in millibars at each reporting station on a surface weather map. Lines called *isobars* are drawn connecting equal values of reported mean sea level pressure. Standard procedure on maps of North America is to draw isobars for every 4 millibars.

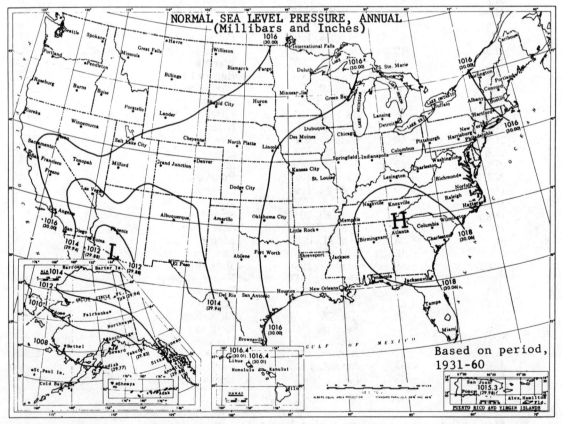

3-6 Normal sea level pressure, annually.

LOWS, HIGHS, AND PRESSURE GRADIENTS

A *low-pressure system* is one in which the barometric pressure decreases toward the center and the windflow around the system is counterclockwise in the Northern Hemisphere. Low-pressure systems with severe storm characteristics are called hurricanes, typhoons, tropical storms, tornadoes, or waterspouts to identify the exact nature of the storm. A *high-pressure system* is one in which the barometric pressure increases toward the center and the windflow around the system is clockwise in the Northern Hemisphere.

The rate of change in pressure in a direction perpendicular to the isobars is called a *pressure gradient*. On a weather map, when the isobars are close together, there is a steep change in atmospheric pressure within a short distance. This can be a sign of rough weather. Figures 3-7 and 3-8 illustrate what a high and a low pressure system would look like if you could see them. The gradient or slope of the pressure isobars is a good indicator of the strength and form of incoming weather. It plays a special part in analyzing and forecasting wind.

WIND

Pressure gradients start the movement of air. As soon as the air builds up speed, the *Coriolis force*, caused by the Earth's rotation, deflects it to the right in the

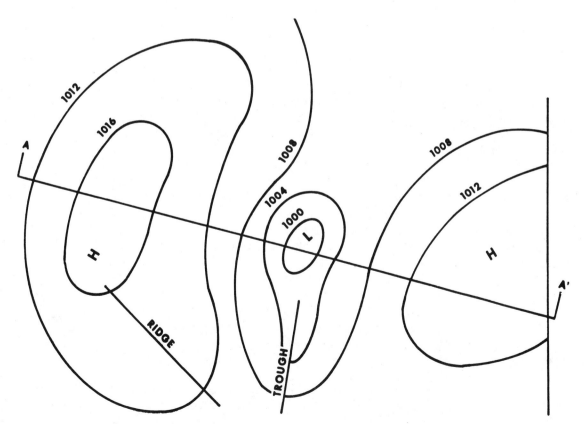

3-7 How high and low-pressure systems look on a map.

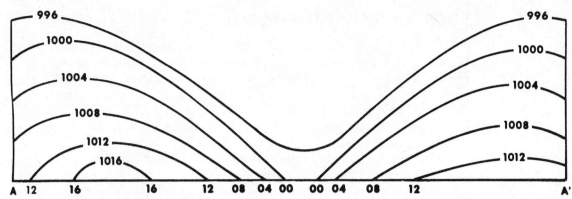

3-8 How the same high and low-pressure systems might look on the ground along line A to A' if you could see the pressure gradients.

Northern Hemisphere or to the left in the Southern Hemisphere. As the speed increases along the isobars, this force becomes equal and opposite to the pressure gradient force and causes what's called the *gradient wind* above 3000 feet.

Friction reduces the surface wind speed to about 40 percent of the speed of the gradient wind and causes the surface wind to flow across the isobars instead of parallel to them. The speed of the surface wind depends on the surface. The friction is least over water and greatest over mountains. The average surface wind will flow across the isobars toward lower pressure at about a 30 degree angle. The surface friction gradually decreases with altitude until the gradient level is reached. Surface winds flow *clockwise* around and away from a center of high-pressure and *counterclockwise* around and toward a center of low-pressure in the Northern Hemisphere (FIG. 3-9).

Other winds come into play to keep meteorologists, professional and amateur, guessing. During daytime, coastal land generally becomes warmer than the nearby water and a lower density will exist in the surface layer of air over land than in the surface layer over the water (FIG. 3-10). This slight difference in pressure over the land and water surfaces starts a flow of wind landward (a sea breeze) during the day. The force of the sea breezes depends on the amount of insolation and radiation. Sea breezes are most pronounced on clear days, in the summer and in low latitudes. Land breezes (from land to sea) occur at night due to the rapid nighttime cooling the land surface (FIG. 3-11).

On warm days, winds flow up slopes during the day and down slopes during the night (FIG. 3-12). This is because air in contact with mountain slopes is warmer than the free atmosphere during the day and colder during the night. Since cold air tends to sink and warm air tends to rise, a system of winds develops and flows up the mountain side during the day and down during the night. The daytime movement is a valley breeze and the nighttime motion is a mountain breeze.

THE JET STREAM

The jet stream is a special circulation and is very important to understanding weather. A *jet stream* is a strong wind within a narrow stream in the atmo-

3-9 The nation's first large experimental windfarm was built at Goodnoe Hills, Washington. This machine was built to produce 2.5 million watts of electricity from the wind's energy (Dept. of Energy).

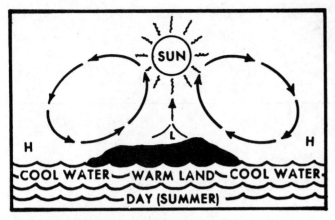
3-10 Land-sea effect during the day.

3-11 Land-sea effect at night.

sphere. While this term can be applied to any such stream regardless of direction, it is commonly interpreted to mean a band or belt of winds that wanders around the globe in a westerly direction. By saying that it is westerly, it is meant that it flows primarily from the west or from adjacent directions such as northwest or southwest. It may range from 25 to 100 miles in width and up to a mile or two in depth. Sometimes the jet stream is a continuous band, but more often it is broken or split at several points.

Jet streams are found in both the Northern Hemisphere and the Southern Hemisphere, but much more is known about the predominant one in the Northern Hemisphere. This is the one normally referred to when only the term "jet stream" is used. It is located in the high tropopause along the boundary of the polar front zone. Normally, there is a break in the tropopause where the jet stream exists, or it can be said that it exists where the tropopause has its greatest slope (FIG. 3-13).

The winds in the jet stream occasionally exceed 250 knots (288 miles an hour). Most of the time the winds range from about 100 to 150 knots (115 to 173

3-12 Effect of windflow over mountains.

miles an hour). However, a band of winds is classed as a jet stream only when the winds in the band have a speed of 50 knots (58 miles an hour) or more. The jet stream is stronger in winter than in summer.

The jet stream is associated with moving low pressure systems and the polar front. It is very important in forecasting weather and predicting the development and movement of fronts and low-pressure systems. It is also important to pilots who fly above 40,000 feet.

MOISTURE

Moisture is a major element of what we call "weather." Moisture causes rain, snow, and humidity. The atmosphere's level and form of moisture can cause comfort or discomfort.

Atmospheric moisture comes in three states: vapor, liquid, and solid. Moisture vapor is called *fog*, liquid is *rain*, and solid is *ice, snow, sleet,* or *hail*, depending on its exact state. Like everything else in weather, moisture must live with certain natural rules, which include evaporation, condensation, sublimation, and heat exchange.

Evaporation is a change in state from a liquid to a gas. As a fun project, set a pan of water out on a warm day and watch it evaporate—change from a liquid to a gas—into the air. It's the same with water in a lake or an ocean. As the temperature of the liquid increases, more of the molecules will escape and the rate of evaporation will increase. This depends on the air pressure and temperature as well as the temperature of the liquid.

Condensation is the reverse of evaporation; it is the change of state from a gas to a liquid. Condensation occurs when the air holding the moisture reaches a saturation point and can no longer hold the moisture in a gaseous form. The moisture is turned into water droplets or, if the air is cold enough, ice. This is what makes rain and snow.

Sublimation is the direct change from solid to vapor and vice versa without passing through the intermediate liquid state. This occurs in clouds at below freezing temperatures as ice and snow form directly from water vapor in the air without first becoming a liquid.

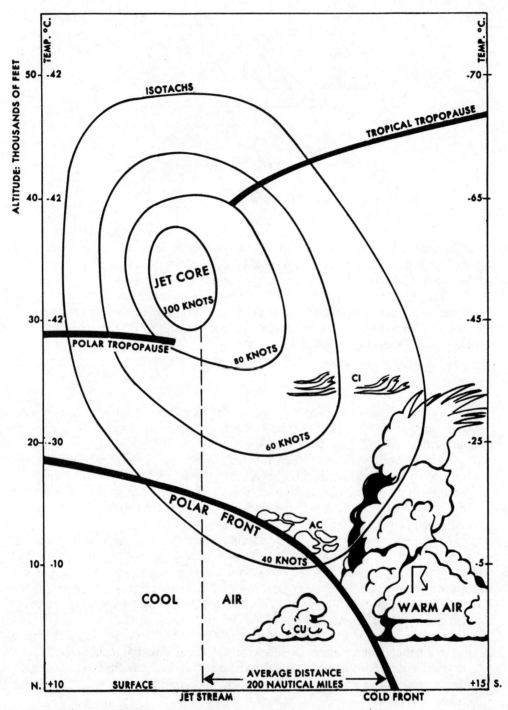

3-13 Vertical cross section of a jet stream.

Heat exchange is the amount of heat needed to change moisture from a gas to a liquid to a solid and back again (FIG. 3-14). Those are the ground rules for moisture. Now let's look at how they react in our atmosphere to give us sunny days and snowstorms.

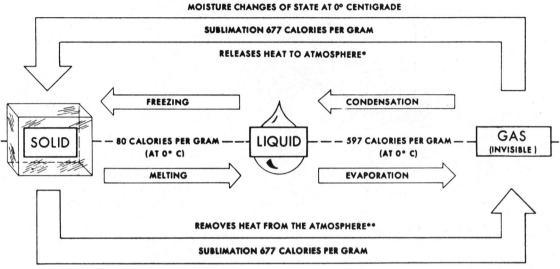

3-14 Heat exchanges.

Saturation occurs when air at a certain temperature and pressure is holding the maximum possible water vapor content. In the atmosphere, saturation occurs by evaporation into the air from a free water surface (ocean, lake, river, rain drops) until vapor pressure is equal between water and air. *Dew point* is the temperature the air must be cooled to become saturated. For example, if the air temperature were 60°F and the dew point 50°F, the air would be saturated if cooled to 50°F. If the air were further cooled to 49°F, it could no longer hold all the water vapor present and some of it would be virtually "squeezed out" in the form of liquid water or condensation (FIG. 3-15).

The moisture content of the air is measured as humidity or the amount of water vapor (in grams) contained in one kilogram of natural air (TABLE 3-5). This is called *specific humidity* (FIG. 3-16). But most of the atmosphere is not saturated. It contains less than the maximum possible quantity of water vapor. For weather analysis it's best to express how near the air is to being saturated. This is the *relative humidity*, the ratio of the amount of water vapor in the air and the amount of water vapor that the air would contain when saturated at the same temperature (FIGS. 3-17 and 3-18). It is written as a percentage, with saturated air having 100 percent relative humidity. It's a handy figure to know.

Precipitation is the general term for all forms of falling moisture—rain, snow, hail, ice pellets, etc. The chain of events that lead up to precipitation include saturation of the air, condensation of moisture into clouds, and heat ex-

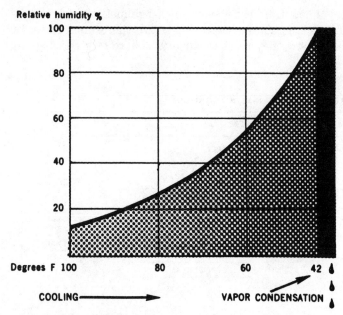

3-15 Relative humidity chart.

Table 3-5 Relative Humidity/Temperature Table.

Temp. °F	Relative humidity %		
	20	30	40
	Equilibrium moisture content (percent)		
40	4.6	6.3	7.9
45	4.6	6.3	7.9
50	4.6	6.3	7.8
55	4.6	6.2	7.8
60	4.6	6.2	7.7
65	4.5	6.1	7.7
70	4.5	6.1	7.6
75	4.5	6.0	7.5
80	4.4	5.9	7.4
85	4.4	5.9	7.3
90	4.3	5.8	7.2
95	4.3	5.7	7.1
100	4.2	5.6	7.0
103	4.1	5.5	6.9
110	4.0	5.4	6.8
115	3.9	5.3	6.7
120	3.8	5.2	6.5

change of cloud vapor into liquid or solids. Precipitation then occurs in the form of liquid or frozen moisture. For average precipitation, see FIG. 3-19.

Rain is precipitation that reaches the earth's surface as relatively large droplets. Rain can be classified as light, moderate, or heavy, based on the rate of fall or the effect it has on surfaces or visibility. Drizzle is precipitation from

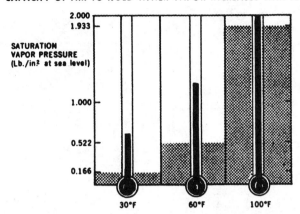

3-16 Capacity of air to hold water is relative to temperature.

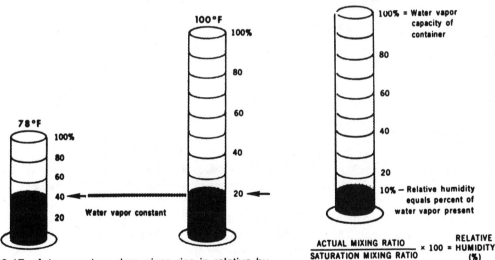

3-17 A temperature drop gives rise in relative humidity.

3-18 Estimating relative humidity.

stratiform clouds in smaller droplets, a sign of little or no turbulence in the air. Drizzle is also classified as light, moderate, or heavy.

Freezing rain or drizzle is precipitation in the form of supercooled liquid raindrops, a portion of which freezes and forms a smooth coating of ice when it strikes the ground. Frozen precipitation comes in three styles: ice pellets or sleet (frozen raindrops formed by rain or drizzle falling through a below-freezing layer of air), hail (lumps or balls of ice circulated through freezing clouds adding layers of ice weighing up to 2 pounds), and snow (ice crystals formed by sublimation in below-freezing clouds).

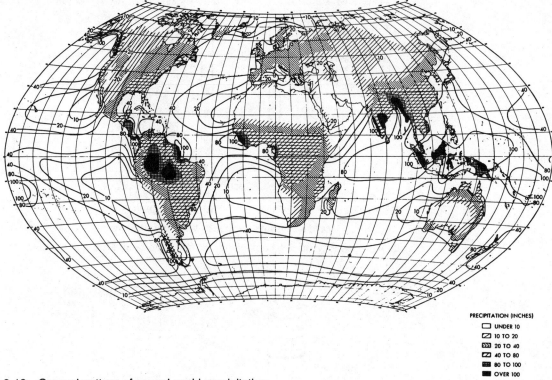

3-19 General pattern of annual world precipitation.

CLOUDS

Cloud formations resulting from saturation-producing processes that take place in the atmosphere. Identifying and interpreting clouds can help you in understanding and forecasting weather.

Clouds are visible collections of condensed moisture consisting of droplets of water or crystals of ice. They are easily supported and transported by air movements as slow as one-tenth of a mile per hour.

The international cloud classification (TABLE 3-6) shows standard cloud symbols. Within this classification, cloud types are usually divided into four major groups and further classified in terms of their forms and appearance. The four major groups are high clouds, middle clouds, low clouds, and vertical clouds.

Within the high, middle, and low cloud groups there are two main subdivisions. Clouds formed when vertical currents carry moist air upward to the condensation level are called *cumuliform-type clouds*, meaning "accumulation" or "heap." These clouds are recognized by their lumpy or billowy appearance. Fliers know that turbulent flying conditions are usually found in and below cumuliform clouds.

Stratiform-type clouds, meaning "spread out", are formed when complete layers of air are cooled until condensation takes place. Stratiform clouds usually look like white sheets.

Table 3-6 International Cloud Classification, Abbreviations, and Weather Map symbols.

Base Altitude	Cloud Type	Abbreviation	Symbol
Bases of high clouds usually above 20,000 feet	Cirrus	Ci	
	Cirrocumulus	Cc	
	Cirrostratus	Cs	
—20,000 feet—			
Bases of middle clouds range from 6,500 feet to 20,000 feet.	Altocumulus	Ac	
	Altostratus	As	
—65,000 feet—			
Bases of low clouds range from surface to 6,500 feet	*Cumulus	Cu	
	*Cumulonimbus	Cb	
	Nimbostratus	Ns	
	Stratocumulus	Sc	
	Stratus	St	
Surface			

*Cumulus and Cumulonimbus are clouds with vertical development. Their base is usually below 6,500 feet but may be slightly higher. The tops of the Cumulonimbus sometimes exceed 60,000 feet.

In addition to the two main subdivisions mentioned, the word *nimbus*, meaning "raincloud," is added to the names of clouds that normally produce heavy precipitation. For example, a stratiform cloud producing precipitation is referred to as *nimbostratus*. A heavy, swelling cumulus cloud that has grown into a thunderstorm is called a *cumulonimbus*. Clouds that are broken into fragment are identified by adding the prefix *fracto* to the classification name. For example, fragmentary cumulus is referred to as fractocumulus.

Cloud Types

The high cloud group includes *cirrus, cirrocumlus*, and *cirrostratus* clouds. The mean base level of these three cloud types is 20,000 feet or higher above terrain. Cirrus clouds show approaching changes in weather. Cirriform clouds are composed of ice crystals. These clouds are usually thin and the outline of the sun or moon may be seen through them, producing a halo effect.

The middle cloud group includes *altocumulus* and *altostratus* clouds. The altocumulus has many variations in appearance and in formation. The altostratus varies mostly in thickness from very thin to several thousand feet. Bases of the middle clouds range from 6500 to 20,000 feet above the terrain. These clouds may be composed of ice crystals or water droplets, and may contain icing conditions hazardous to aircraft. Altocumulus rarely produces precipitation, but altostratus usually show that unfavorable weather and precipitation are nearby.

The low cloud group consists of *stratus, stratocumulus*, and *nimbostratus* clouds. The bases of these clouds range from near the surface to about 6500 feet above the terrain. The heights of the cloud bases can change rapidly. If low clouds form below 50 feet, they are called fog and can completely blanket landmarks. Low clouds have the same components as middle clouds.

Clouds with vertical development include the cumulus and cumulonimbus clouds. These clouds generally have their bases below 6500 feet above the terrain and tops sometimes exceed 60,000 feet. Clouds with vertical development are caused by some type of lifting action, such as convective currents, convergence, orographic lift, or frontal lift.

The cumulonimbus cloud is what is commonly called the thunderstorm cloud. Thunderstorms are almost always accompanied by strong gusts of wind and severe turbulence, heavy rain showers, lightning, and other surprises. Frontal thunderstorms are caused by the lifting of warm, moist, conditionally unstable air over a frontal surface. Thunderstorms might also occur many miles ahead of a rapidly moving cold front. Air mass thunderstorms are contained within air masses and can occur after a front has moved through. Figures 3-20 through 3-22 show the stages of thunderstorm development.

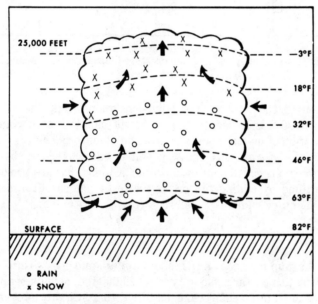

3-20 Cumulus stage of a thunderstorm cell.

Tornadoes And Waterspouts

A *tornado* is a very violent whirling storm with a small diameter, usually a quarter of a mile or less. The length of the track of a tornado on the ground may be from a few hundred feet to 300 miles. The average is less than 25 miles. When not touching the ground it is called a *funnel cloud*. Wind velocity within a tornado usually range between 150 and 300 miles an hour while it tracks along the ground at a comparatively slow 25 to 40 miles an hour.

Most tornadoes in the United States occur in the late spring and early summer and are usually found near thunderstorm activity and heavy rains (FIG. 3-23).

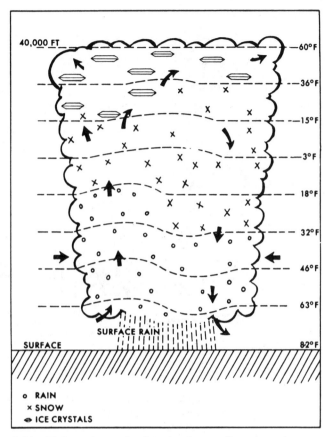

3-21 Mature stage of a thunderstorm cell.

The majority of tornadoes appear about 75 to 180 miles ahead of a cold front along the prefrontal squall line (FIG. 3-24).

Waterspouts are tornadoes that form over ocean areas. One type is a true waterspout in which the funnel forms at the cloud and extends to the surface, usually in advance of a squall line.

THE HYDROLOGIC CYCLE

Water at the Earth's surface evaporates by absorbing heat energy. The water vapor is moved by atmospheric currents until it releases its energy to the atmosphere in the form of clouds and fog. The water cycle is completed when precipitation returns the water to the surface. This evaporation-condensation cycle is called the *hydrologic cycle*.

Man's desire to understand this cycle of life is not new. The Biblical Book of Job says "He (God) draws up the drops of water, which distill as rain to the streams; the clouds pour down their moisture and abundant showers fall on mankind. Who can understand how he spreads out the clouds, how he thunders from his pavilion?" (Job 36:27-29–New International Version).

62 THE ELEMENTS

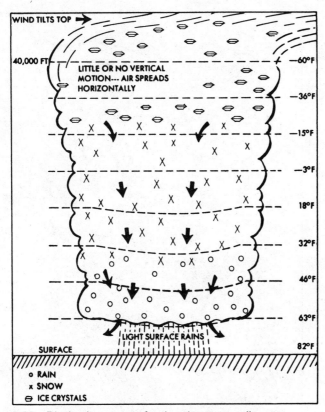

3-22 Dissipating stage of a thunderstorm cell.

Our understanding of the miracles of the atmosphere is still small, but, with knowledge from observation, we can become weather prophets.

(For climatic records from around the world see TABLE 3-7.)

DO YOU KNOW...

- The six elements that make up the weather?
- How to convert Celsius temperatures to Fahrenheit?
- What causes the temperature to change?
- The standard atmospheric pressure at sea level?
- Where the jet stream is?
- What the *hydrologic cycle* is?

Do You Know...... 63

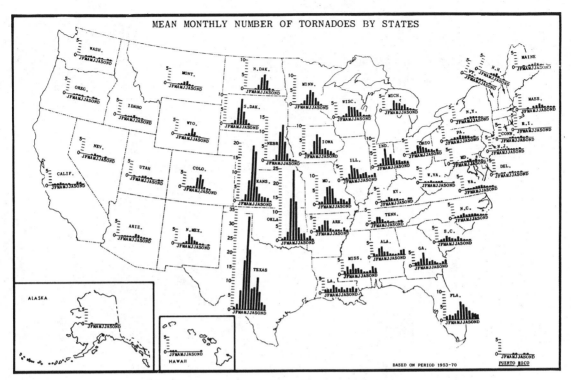

3-23 Mean monthly number of tornadoes by state.

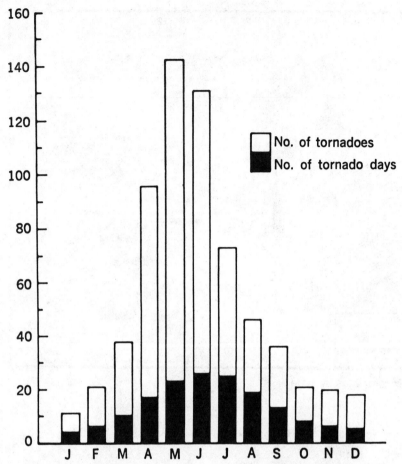
3-24 Average monthly number of tornado days in the United States.

Table 3-7 Worldwide Temperature and Average Precipitation Extremes

Key No.	Area	Highest °F	Place	Elevation Feet	Date
1	Africa	136	Azizia, Libya	380	Sept. 13, 1922
2	North America	134	Death Valley, Calif.	-178	July 10, 1913
3	Asia	129	Tirat Tsvi, Israel	-722	June 21, 1942
4	Australia	128	Cloncurry, Queensland	622	Jan. 16, 1889
5	Europe	122	Seville, Spain	26	Aug. 4, 1881
6	South America	120	Rivadavia, Argentina	676	Dec. 11, 1905
7	Oceania	108	Tuguegarao, Philippines	72	Apr. 29, 1912
8	Antarctica	58	Esperanza, Palmer Pen.	26	Oct. 20, 1956

Key No.	Area	Lowest °F	Place	Elevation Feet	Date
9	Antarctica	-127	Vostok	11,220	Aug. 24, 1960
10	Asia	-90	Oymykon, U.S.S.R.	2,625	Feb. 6, 1933
11	Greenland	-87	Northice	7,690	Jan. 9, 1954
12	North America	-81	Snag, Yukon, Canada	1,925	Feb. 3, 1947
13	Europe	-67	Ust'Shchugor, USSR	279	January+
14	South America	-27	Sarmiento, Argentina	879	June 1, 1907
15	Africa	-11	Ifrane, Morocco	5,364	Feb. 11, 1935
16	Australia	-8	Charlotte Pass, N.S.W.	...	July 22, 1947*
17	Oceania	14	Haleakala Summit, Maui	9,750	Jan. 2, 1961

+ exact date unknown; lowest in 15-year period
* an earlier date
... elevation unknown

Key No.	Area	Greatest Amount Inches	Place	Elevation Feet	Years of Record
18	Oceania	460.0	Mt. Waialeale, Kauai, Hawaii	5,075	32
19	Asia	450.0	Cherrapunji, India	4,309	74
20	Africa	404.6	Debundscha, Cameroon	30	32
21	South America	353.9	Quibdo, Colombia	240	10–16
22	North America	262.1	Henderson Lake, B.C., Canada	12	14
23	Europe	182.8	Crkvica, Yugoslavia	3,337	22
24	Australia	179.3	Tully, Queensland		31

Key No.	Area	Least Amount Inches	Place	Elevation Feet	Years of Record
25	South America	0.03	Arica, Chile	95	59
26	Africa	< 0.1	Wadi Halfa, Sudan	410	39
27	Antarctica	*0.8	South Pole Station	9,186	10
28	North America	1.2	Batagues, Mexico	16	14
29	Asia	1.8	Aden, Arabia	22	50
30	Australia	4.05	Mulka, South Australia		34
31	Europe	6.4	Astrakhan, USSR	45	25
32	Oceania	8.93	Puako, Hawaii	5	13

* The value given is the average amount of solid snow accumulating in one year as indicated by snow markers. The liquid content of the snow is undetermined.

4
The Weather Eye

DO-IT-YOURSELF WEATHER FORECASTS HAVE BEEN MADE THROUGHOUT THE AGES. AS EARLY as 400 B.C. the Greek philosopher Aristotle wrote *Meteorologica* dealing with the "study of things lifted up."

Early forecasts were based on local observations made by people. Accurate measurements of temperature and atmospheric pressure were not available until after the thermometer and the barometer were developed in the 17th century. Weather forecasting did not become practical until the telegraph was invented in the 19th century. This made it easier to collect and transmit weather information from many locations.

The first systematic weather observations in the United States began in 1738. One of the world's first known weather maps was drawn by the German scientist Heinrich Brandes in 1816. The next step was a telegraphic network of observers needed to prepare daily weather maps. This came about in 1849 through the efforts of Joseph Henry of the Smithsonian Institute in Washington, D.C.

The National Weather Service has its roots in the United States government's first official weather forecasts, offered in 1870 through the efforts of the Army Signal Service, which controlled the government's telegraph system. In 1891, the Army's civilian weather activities were moved to a new agency called the United States Weather Bureau, within the Department of Agriculture. In 1940, the Weather Bureau moved in with the Department of Commerce. In 1970, it was renamed the National Weather Service and placed under the Department of Commerce's National Oceanic and Atmospheric Administration (NOAA).

The *Canadian Meteorological Service* was set up in 1871. Exactly a century later it was renamed the Atmospheric Environment Service and became agency of the Department of the Environment.

There's even a *World Meteorological Organization* (WMO), established in 1951 by the United Nations. It coordinates the worldwide exchange of weather and climate information.

THE NATIONAL WEATHER SERVICE

The National Weather Service (NWS) has a large network. NWS personnel are found at over 400 facilities in the 50 states and elsewhere. The NWS has about 5000 full-time employees working in meteorological, hydrological, and oceanographic operations. In a single year, about 3.5 million observations are taken and 2 million forecasts and warnings are issued. In addition, millions of individual briefings and services are provided on a routine but unscheduled basis to aviators, boat and ship operators, and the general public.

The National Meteorological Center (NMC) is the operating nerve center for weather information. The Center's computers incorporate more than 100,000 weather reports daily from around the world into physical and numerical models of the atmosphere. These produce weather predictions as far as ten days into the future, as well as monthly and seasonal predictions of expected temperature and precipitation conditions over North America.

Weather Service Forecast Offices (WSFOs) issue warnings and forecasts for over 600 zones throughout the U.S. The forecasts are issued three times a day for 48 hour periods and are updated as necessary. Extended forecasts, looking ahead five days, are issued daily for statewide areas. Local forecasts, adaptations of zone forecasts for metropolitan areas, cities, and towns, are issued by 52 forecast offices and more than 240 smaller Weather Service Offices.

Even with all their sophisticated computers and charts, the National Weather Service can be wrong. They just forecast the weather; they don't make the weather. Many times the weather observations of the human eye are nearly as accurate as technology.

Let's look back at what mankind knew about forecasting weather before science came along.

WEATHER PROVERBS

The trouble with weather proverbs is not that they are all wrong, but that they are not all right for all times in all places. Some proverbs heard in New England originated thousands of years ago in northern Africa near the Mediterranean Sea, where they were heard and repeated and at last recorded by the writers of the Old Testament. Many a farmer in the Midwest, depending on a sure-fire weather saying his grandfather brought from Germany or Sweden, found it useless in the United States.

Distances far shorter than in either of these examples are enough to ruin some weather proverbs. Those, for example, that predict rain from the direction on the wind. When the wind blows up the side of a mountain it is cooled and loses its moisture in the form of rain. A west wind blowing up the west side of a mountain would produce the same result, a fall of rain, as an east wind blowing up the east side of the same mountain. What this adds up to is that a distance just great enough to hold a good sized mountain might also be great enough to ruin a proverb about west (or east) winds bringing rain.

Here are a few contradictory weather proverbs that were no doubt written in different places:

"Fair weather cometh out of the north." (Job)

"The north wind bringeth forth rain." (Proverbs)

"Take care not to sow in a north wind or to graft and inoculate when the wind is in the south." (Pliny)

"The north wind is best for sowing seed, the south for grafting." (Worledge, 1669)

Another point worth noting about the importance of location is that our Pacific Coast offers moisture-bearing winds from the west and southwest, while in the East they come from over the Gulf of Mexico and the Atlantic. These two weather proverbs, then, should not be considered too seriously in the East:

"A western wind carrieth water in his hand."

"When the east wind toucheth it, it shall wither."

On the other hand, few people on the west slopes of the Cascade Mountains and the Sierras where rain and snow are very frequent companions of west and southwest winds would agree with the following proverb:

"When the wind is in the west, the weather is always best."

Proverbs about the southerly breezes can also be misleading, depending on where you live:

"The south wind warms the aged," and

"The south wind is the father of the poor," are hardly applicable in Louisiana. South winds are about the wettest, stormiest and generally least pleasant of winds in states bordering the Gulf of Mexico. The proverb writers, including Shakespeare, pointed this out:

"The south wind doth play the trumpet to his purposes, and by his hollow whistling in the leaves foretells a tempest and blustering sky."

"If feet swell, the change will be to the south, and the same is a sign of a hurricane."

"When the wind's in the south the rain's in its mouth."

Anybody who looks at a collection of these sayings must be impressed by their variety. They are extremely ancient, about as old as language itself. They illustrate the importance of weather in human affairs. They show man's hopeful opinion that experience is a good teacher. They produce some very striking relationships: wolves and crops, sky colors with foul results, holy days and unholy weather.

Rain is foretold by the behavior of cats, dogs, cattle, red hair and ropes, spiders and smoke, crickets, frogs, birds, mice, flies, and rheumatism. Squirrel stores and the thickness of their fur lead to prophesies of hard winters. The drought or wetness of summers is predicted by the weather in March. What happens on Christmas foretells what will happen on Easter; light or heavy fogs in October foretell light or heavy snows in the coming winter. One proverb says "If the spring is cold and wet, then the autumn will be hot and dry," another says "A wet fall indicates a cold and early winter," and still another "A cow year is a sad year and a bull year is a glad year."

A few other weather proverbs: good, bad, and indifferent illustrate the variety of forecasting opinions:

"When the wind is in the south it blows the bait in the fishes' mouth."

"One swallow does not make a summer."

"If the weather is fine, put on your cloak; if it is wet, do as you please."

"A bad year comes in swimming."

"The first Sunday after Easter settles the weather for the whole summer."

"If Candlemas Day be fair and bright, winter will have another flight; but if Candlemas Day brings clouds and rain, winter is gone and won't come again."

"Red sky in the morning, sailors take warning."

"Mare's tails and mackerel scales make tall ships take in their sails."

"March comes in like a lion and goes out like a lamb."

"Clear Moon, frost soon."

"Rain before seven, stop by eleven."

"A year of snow, a year of plenty."

"Rainbow in the morning gives you fair warning."

"The bonnie Moon is on her back, mend your shoes and sort your thack."

"When the stars begin to huddle, the Earth will soon become a puddle."

"A windy May makes a fair year."

"Do business with men when the wind is in the northwest."

One of the best known of the rain prophesies states that 40 wet days will follow a rainy St. Swithen's Day (July 15). The Groundhog Day story gets into practically every newspaper in the country during the first week in February.

Since neither of these old standbys has any basis in meteorology their popularity, like that of countless others, must be explained by something else. Nearly everybody on Earth since Creation has wanted to know what the weather is going to be tomorrow, next week, next month, and a year from now.

Farmers and sailors want to forecast weather because a large part of their work and fortunes depend on the weather. But weather also affects many others: salesmen, grain speculators, baseball players, amusement park managers, fishermen, military leaders, and everyone else from advertisers to zookeepers.

The best explanation for the spread of these weather sayings is simply that a great many of them make good sense. For example, "One would rather see a wolf in February than a peasant in his shirtsleeves," means simply that a warm February will advance plant growth so far that a hard frost will destroy it, which nobody wants, especially a farmer who depends on his crops for a livelihood. Here are three others with the same message:

"A late spring never deceives."

"Better to be bitten by a snake than the feel the Sun in March."

"A wet March makes a sad harvest."

The familiar halo of the Sun or Moon is caused by the refraction of its light by ice crystals in cirrus clouds, which frequently appear when lowered air pressure and high clouds are present and rain is approaching. Thus, proverbs saying the ring around the Sun (or Moon) is a sign of rain, such as "the Moon with a circle brings water in her beak", are frequently right.

Several of the many signs seen in the behavior of animals and insects are worth noting, too. For example:

"A bee was never caught in a shower."

"Expect storm weather when ants travel in lines, and fair weather when they scatter."

"When flies congregate in swarms, rain follows soon."

"Pigeons return home unusually early before rain."

The following proverb gives several results of the low air pressure or high humidity that often warn us of rain should prove, if we wait long enough, that not all weather signs are wrong.

"Lamp wicks crackle, candles burn dim, soot falls down, smoke descends, walls and pavements are damp, and disagreeable odors arise from ditches and gutters before rain."

Finally, this weather proverb of dubious meteorological value:

"Dirty days hath September
April, June and November;
From January up to May
The rain it raineth every day.
All the rest have thirty-one

Without a blessed gleam of Sun;
And if any of them had two and thirty;
They'd be just as wet and twice as dirty."

THE HUMAN CONDITION

One of the best ways of observing weather firsthand is by observing yourself. That's right, your body responds to changes in air pressure, humidity, and temperature. Reading the signs can give you a hint of coming weather and can improve your health.

In fact, there's a field of medical science called *biometeorology* that studies the relationship between weather and health. Medical studies have illustrated that people respond to almost every shift of the changing atmosphere that covers the Earth. Body changes include pulse rate, body temperature, blood pressure, and urine. Individual reactions to the environment differ, just as individuals are physically different.

Sudden, severe changes in the weather have the greatest impact on people who suffer from arthritis, headaches, asthma, heart disease, or mental illness. Modern living has made it more difficult for us to adapt to the elements. Societies that live in nature; away from air conditioning, humidifiers, smog, and other unnatural conditions adapt more easily to changing weather conditions.

Hippocrates, the "father of medicine," wrote that hot and cold winds have a bearing on the outbreak and course of disease. But he didn't know why. Some winds carry fungi spores that irritate people with asthma. Other winds mark pressure changes that irritate health.

Some people are sensitive to the electromagnetic waves that come from thunderstorms and can "feel it coming" miles away. Increasing barometric pressure can sometimes be felt by the blood, veins, and heart, lowering oxygen levels and adversely affecting people with angina. When the barometer falls; some people feel edgy; others feel happy. It's been proven that during these times crime rates rise slightly.

Moist days are healthier than dry days. Extreme dryness can cause irritation of the nose and throat, encouraging many viruses to spread. This is why doctors suggest humidifiers in drier climates.

Your build also tells you something about the weather. If you are thin, not well insulated by fat, you may be more conscious of cold because you store less energy and water. Shifting weather can more easily be felt by your body. Obese people are better insulated against winter's cold than those with heavy muscles.

Cold winters increase the chances of colds, not because of an increase of viruses as much as the increased contact with people in crowded places.

Let's consider how weather affects specific ailments.

Arthritis: The combination of rising humidity and falling barometric pressure causes arthritic joints to ache because albuminlike fluid increases the size of joint tissues and causes resistance to motion and pain of movement.

Asthma: Doctors aren't sure why, but dramatic changes in the weather affect asthma patients. The greatest culprit is winds that carry pollens and other substances that irritate the lungs of sensitive people. The first invasion of cold air each year can also initiate asthmatic spasms.

Headaches: Atmospheric pressure changes and intense sunlight are the causes of many headaches according to doctors. As most headaches are within the sinus passages, low humidity in the air and sharp pressure changes are a shock to the sinus and cause sinus headaches, some of them severe.

Common cold: Weather conditions don't cause colds, but they can make them easier to catch and harder to get rid of. The cold viruses and bacteria are always present in your body but sharp changes in the weather can reduce your body's ability to fend them off.

Bronchitis and Emphysema: Victims of these diseases need oxygen and anything that robs the air of oxygen can make breathing more difficult and spur an attack. Weather-related causes include inversion layers over metropolitan areas that stagnate air and pollutants. High humidity also lessens the amount of oxygen in each breath and makes patients uncomfortable.

Heart attack: Research shows that sudden changes in weather, such as fast-moving fronts, changes in barometric pressure, and moisture can trigger heart attacks. So can the unfamiliar exertion of shoveling new snow.

WEATHER AND THE MIND

Weather can also affect people's minds. Moods can change, irritability can increase, crime can multiply—all because of changes in the weather.

Some people are very sensitive to weather and respond to every cloudy and every sunny day. Their moods also reflect changes in barometric pressure and humidity. Ideal weather conditions for humans are an average temperature of 64°F with about 65 percent humidity, day and night. Some can adapt easily to sudden changes from this norm but others feel uncomfortable when conditions move far ahead.

Sharp changes in atmospheric pressure, studies say, can trigger more aggressive behavior. A Canadian study found that most car accidents occur when the barometer is falling. A study in Tokyo found that falling pressure meant that the lost and found offices were unusually busy; people were more absentminded. Suicide attempts seem to be more likely when the barometer is falling.

Another study found that assaults and related crimes increased by nearly half on warm, muggy days and nights. Conversely, seasonably comfortable weather after a period of bad weather often drops the crime rate by 75 percent.

One excellent way of observing and forecasting weather is to watch your own body react to changes in the weather. It will not only give you a clue to incoming weather conditions, it will also help you understand and take advantage of physical and mental changes caused by the weather.

SKY OBSERVATIONS

You possess the greatest weather eye of all. You can see and interpret the changes in the weather around you: clouds, precipitation, winds, and nature.

Here are some practical weather observations you can make to help you foretell future weather conditions.

Storm Indicators

Nature reacts to incoming storms in various ways: insects are more active, bees return to hives, bats and birds fly lower, and frogs croak more, all in response to changing atmospheric pressure and humidity.

Indications that precipitation, rain or snow, is on the way include: low clouds moving in behind middle and high clouds, a ring around the moon (indicating high cirrus clouds), clouds begin developing vertically, a dark threatening sky is seen to the west or northwest (in most areas), or when leaves show their undersides (due to strong winds that often precede cold fronts).

Blue Skies

Fair weather can be forecast as dark clouds becoming lighter and steadily rising in altitude (because of the passing of the cold front), a sudden wind shift (meaning the passing of the cold front), and warming winds from the direction of prevailing warm fronts in your area.

Dew is also an indicator of clear weather.

Other Weather Phenomena

Frost can often be foretold by fog forming on a pond in the spring or fall, indicating colder air temperatures that turn moist air vapor into water.

KEEPING TRACK

Many people keep track of weather observations in a weather diary or as regular entries in their personal diary or journal:

> "September 7, noted a red halo around the Moon tonight which could mean rain tomorrow."

> "The animals seem more active today. There must be a low front coming in. Watch for storm on the horizon."

Once you've set up your own weather observation station with the instruments discussed in the next two chapters, you can make more specific entries in a weather log shown in Chapter 7.

In the meantime, you may come up with some of your own weather proverbs developed with your region in mind. Watch for the direction of incoming weather, how it is affected by mountains and bodies of water in your region, how long systems take to move from the horizon to your home, and how changing weather conditions affect your mind and your body.

You can be your own weather eye.

DO YOU KNOW...
- How the National Weather Service started?
- Why weather proverbs are not always true?
- How the weather can affect your mind and body?
- How the weather can cause headaches?
- How some people can "feel" a thunderstorm coming?
- How to forecast a storm with your eyes?

5
Basic Instruments

IF YOU'VE EVER SAID, "IT LOOKS LIKE IT'S GOING TO RAIN," YOU'VE PRACTICED METEOROLOGY and, on a small scale, imitated the National Weather Service's major jobs: observing, forecasting, and distributing weather information.

A number of people, however, are more than just casual observers of the passing weather scene. Many have set up their own stations and have fun recording each day's weather and trying their hand at forecasting. These people are called amateur meteorologists, and you can join them. With some basic instructions and a few simple instruments, some of which you can build at home, you can be an amateur weather forecaster (FIG. 5-1).

Observing, recording, and forecasting are the present, past, and future tenses of meteorology.

Observations, taken accurately and at regular intervals, are important to the weather forecaster. Weather changes cannot usually be predicted more than a few hours in advance. Indeed, many local storms give scarcely an hour's notice of their arrival.

Records kept of past weather or climate can be a valuable tool in predicting the weather. The rules in this book for forecasting weather from local readings are usually reliable, but may not apply in all locations and situations. To form your own rules for your area you must carefully record and review your own observations.

Forecasting is the real challenge of meteorology. Sailors are said to be able to forecast weather changes from local observations. Pilots and farmers who depend more on the weather also become amateur meteorologists. This is because they must depend on the weather for their jobs and their safety.

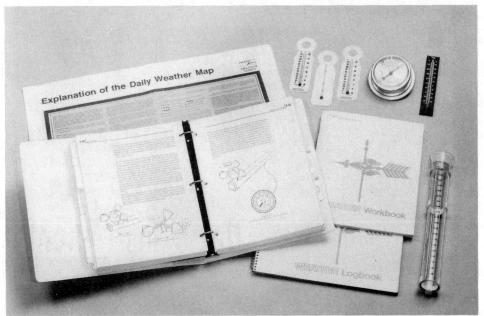

5-1 Basic weather instruments are important to studying meteorology. (Courtesy Heath Company)

THE WEATHER SHACK

The first thing you'll need at your weather station is a *weather shack*, a small shelter for the instruments. For best results, thermometers should be kept outdoors in a shelter. A thermometer really only takes its own temperature and it is to measure the air temperature the thermometer must be exposed to the air. This means keeping it out of the Sun and keeping it where the air can move freely around it. The best place for it is a ventilated box or thermometer shelter.

You can build your own weather shack from a couple of window shutters, some plywood, and a supporting post or posts (FIG. 5-2). The shutters form the sides of the shelter. The back, top, and bottom can be made from two layers of wood with a small air space in between. It can have a hinged door or an open side, but the opening should face north to keep the Sun from shining in and affecting the thermometers when the door is open. The shelter should be painted white, both inside and out.

WIND INSTRUMENTS

Wind is air in motion. As air in motion, the wind has four properties important to the meteorologist: direction, speed, character, and shifts.

Wind *direction* is the direction from which the wind is blowing. It is usually expressed in compass points, northeast, southwest, east, or in degrees from north 180 (south), 230 (southwest), etc.

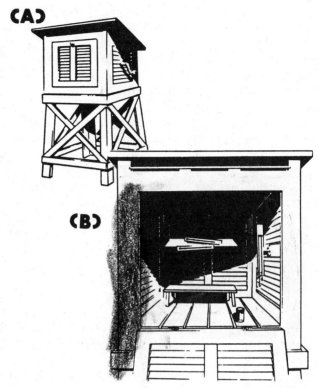

5-2 The standard instrument shelter or "weather shack," (A) how it's built and (B) what it contains.

Wind *speed* is the rate of motion of the air in a unit of time. Wind speed can be measured in *knots* (nautical miles per hour) or *mph* (miles per hour).

The *character* of the wind refers to its gustiness. A *gust* is defined as a rapid fluctuation in wind speed with a variation of 10 knots or more between peaks and lulls. *Squalls* are sudden increases in wind speed of at least 15 knots sustained at 20 knots or more for at least one minute. *Peak gust* is the highest instantaneous wind speed seen or recorded.

Wind *shift* is a change in wind direction of 45 degrees or more that takes place in less than 15 minutes. Wind shifts normally occur with the passage of a cold front. Wind shifts foretell of a rapid drop in the temperature and dew point, a rapid rise in pressure, and a coming storm.

Wind Vane

Wind *direction* can be estimated from the flow of smoke, the behavior of a flag, or even the popular wet-finger method: wet the end of your finger with your mouth and hold your finger in the air to feel the wind. If a wind vane is in sight, so much the better.

A wind vane can be easily made from wood, tin, or other material. Both ends should be balanced and it should swing freely on a central pivot. Decorative wind vanes can be purchased at small cost, adding a decorative value to their scientific usefulness.

The vane should be mounted where the wind is least affected by local influences such as trees, buildings, and windbreaks. The direction indicators should be carefully lined up with the compass points.

You can also make a wind sock to give you both wind direction and its approximate speed. A square yard of light-weight material can be cut in a wide funnel shape, 36 inches wide narrowing down to 12 inches. The fabric is then sewn together to form a wind funnel and attached to a piece of heavy wire or a coathanger formed in a circle. Curtain rings will keep the sock loosely attached to the frame. A cardboard circle can then be cut out, marked with the compass points, and slid onto the broomstick or garden stake that will hold the windsock in place. A compass can align the cardboard marker. Mounted clear of obstructions, your wind sock can show you the direction from which the wind blows as well as estimated speed. A limp windsock means calm, straight out means about 30 knots (35 mph) and halfway between means a wind of about 15 knots (17 mph).

Anemometer

To measure wind speed more accurately you'll need a spinning *anemometer*. They can be built at home, but few are as satisfactory as a cup-type anemometer with a gear box.

One useful homemade anemometer is shaped somewhat like a slingshot with a flap suspended between its arms which swings back along a scale as the wind strikes it. This instrument is tested by holding it out of the window of a moving car and marking the point to which the flap is blown back along the scale at different speeds.

As with the wind vane, readings with this instrument should be made where the wind is least affected by local influences. The anemometer should be headed directly into the wind while a reading is being made.

BAROMETER

While the barometer has been greatly overrated as a forecasting device, it is important that the local observer be equipped with some means for detecting changes in atmospheric pressure. That's the job of the barometer. Pressure changes used in connection with the wind direction will give you the best key to the local weather situation (FIG. 5-3).

The barometer is, by far, the most expensive instrument the amateur meteorologist might have to purchase. Homemade barometers are rarely accurate enough for serious weather observation. If your budget is small, you can use a weather radio to learn the latest barometric pressure in your area. Otherwise, a serviceable aneroid barometer can be purchased for about $40.

The barometer should not be placed out-of-doors because such exposure can corrode its parts. Since air pressure is the same both indoors and outdoors, the instrument can be located inside in some convenient spot. Keep it out of sunlight

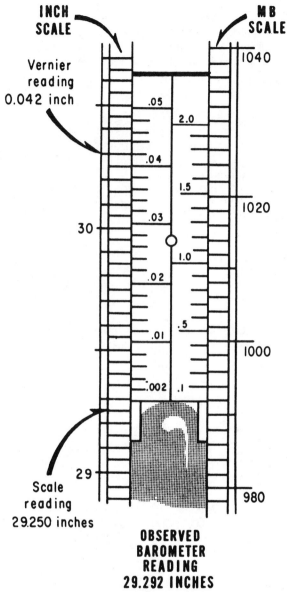

5-3 Reading the mercurial barometer in inches of mercury and millibars.

and away from drafts where the temperature does not vary too much and the readings will be reliable.

The barometer should be adjusted to sea level. Instructions for such adjustments will be included with the barometer. The current sea level pressure is available from local weather service facilities.

Of the instruments combined in thermometer-barometer-humidity instrument sets, the barometer will be the only aid to the amateur meteorologist. Temperature and humidity readings taken indoors are worthless in forecast work and only give you conditions in the house. Better quality instrument sets remotely read outside temperatures and humidity.

To help you in shopping for the best barometer for your weather station, let's look at the types, how they work, and what they do.

Mercury Barometers

Nearly three and a half centuries ago, an Italian physicist named Torricelli made the first crude barometer. He used a long glass tube, open at one end, closed at the other, and filled with mercury. The open end was sealed temporarily and placed into a basin of mercury, then unsealed (FIG. 5-4). This allowed the mercury in the tube to descend, leaving a nearly perfect vacuum at the top of the closed end of the tube.

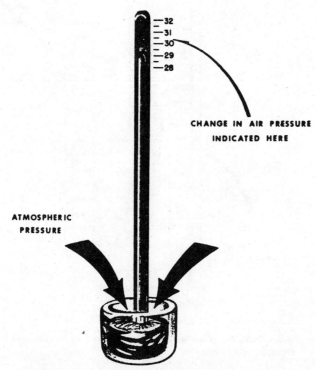

5-4 Principle of the mercurial barometer.

When the atmospheric pressure is increased the mercury in the basin is forced into the glass tube. As the atmospheric pressure is decreased, the mercury in the tube flows into the bottom. The height of the mercury column in the tube becomes a measure of the air pressure. This is where we get the pressure measurement term *inches of mercury*.

Aneroid Barometer

Mercury barometers are quite accurate, but they are expensive and not easy to move around because they are nearly three feet high. For most purposes they are replaced by a mechanical instrument known as the *aneroid barometer*.

The aneroid barometer uses the change in shape of an evacuated metal cell to measure changes in atmospheric pressure. The aneroid barometer gets its name from the pressure-sensitive element used in the instrument. It is an *aneroid*, which is a thin walled metal capsule or cell, sometimes called a diaphragm, that has been either partially or completely evacuated of air, a vacuum. The aneroid is usually made of beryllium copper or phosphor bronze. In a common type of single aneroid cell barometer (FIG. 5-5), the top of the evacuated cell is secured to a linkage which transfers the aneroid's motion to a pointer. The pointer reads the pressure on the face of the instrument.

5-5 Simple diagram of the aneroid barometer.

The best barometer is a computerized barometer that uses an electronic transducer to read atmospheric pressure. One such model offers current pressure, rate of change, direction of change, maximum and minimum pressures from memory, as well as the date and time they occurred. It is also a clock. Readings are given in inches of mercury, millibars, or kiloPascals (ten millibars).

THERMOMETERS

The thermometer is another important weather instrument that can bought for a few dollars.

If maximum and minimum temperature records are to be kept, a special thermometer or set of thermometers will be needed. The National Weather Service uses a set of two thermometers to obtain maximum and minimum readings. The maximum thermometer has a special construction near the bulb and the minimum thermometer has a floating index.

A less expensive, longtime favorite of the amateur forecaster is called the *Six's-type*, name after its inventor. This U-shaped thermometer with one side reading maximum and one side reading minimum temperatures can be purchased from some hobby shops for less than $25.

A dial type maximum and minimum thermometer is also available. This type has a coiled metal spring temperature element and two thin metal pointers that are pushed either up or down into position as the main point of the temperature element moves around the scale.

Thermometers should be located outdoors in the shade. An instrument shelter is ideal for thermometer exposure. If a shelter is not available, thermometers can be mounted on the shady side of unheated buildings or on trees.

The standard air thermometer that you are familiar with (FIG. 5-6) is the one placed inside or outside the house to see how cold or warm the temperature is during the day or how cool the air conditioner is keeping the house. There are many uses for this thermometer, which is usually filled with mercury or alcohol. These two fluids are used because they expand with temperature than does the glass container (FIG. 5-7). The standard air thermometer used by most stations has a range from −20°F to +120°F.

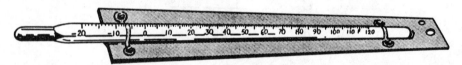

5-6 Standard air thermometer.

Handle these thermometers carefully because they break easily. It is important that the thermometer stem and bulb be kept clean and free of dirt, dust, and moisture since these elements can cause error in free air temperature readings. Clean the stem and bulb by wiping with a soft cloth. This should be done 10 to 15 minutes prior to taking a reading so the temperature will have time to stabilize before the observation. Remove and clean the metal back as necessary. Upon reassembly apply a drop of light oil to any mounting screws, if needed. Renew the etched graduations when faded.

If the mercury in your thermometer has a bubble or open space in it, tightly attach a string to the thermometer back and whirl it. You can also tap the bulb lightly against the heel of the fleshy part of your hand to jar the mercury column back together. If this fails, gently heat the bulb by placing it near a light bulb until the column unites. Never heat the bulb over an open flame. Leave a small space at the top of the tube while heating; otherwise the thermometer will break. If these methods fail to unite the mercury column, the thermometer should be replaced.

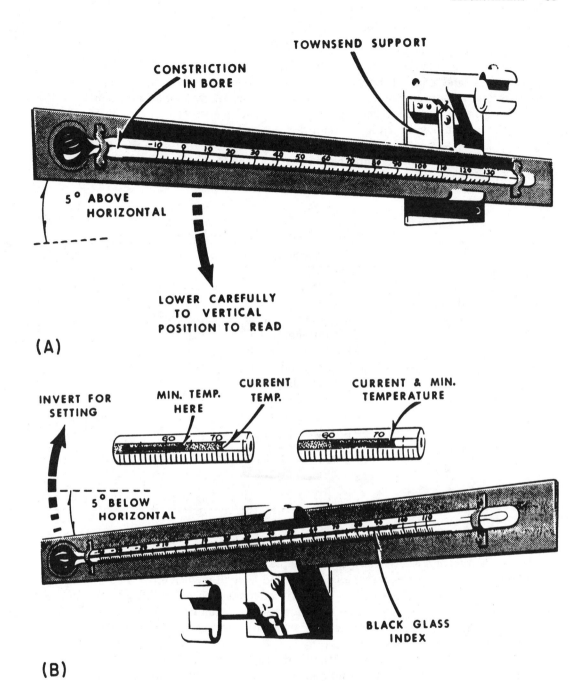

5-7 Maximum thermometer (top) and minimum thermometer (bottom).

HYGROMETERS

There are six different types of *hygrometers* to measure the water vapor content (humidity) of the atmosphere. The psychrometer is the simplest type and can be built at home without too much trouble or purchased at a reasonable cost.

The *psychrometer* is a set of two matching thermometers. One, exposed to the free air, is the dry bulb and the other has its bulb covered by a water-saturated wick, and is called the wet bulb. The wet bulb measures the temperature at which water is evaporating from the wick. Since the rate of evaporation is controlled by the amount of moisture in the air and evaporation is a cooling process, the wet bulb will read lower than the dry bulb. This difference in temperature is known as the *wet bulb depression*. When the wet bulb temperature is subtracted from the dry bulb temperature, TABLE 5-1 can be used to find out the relative humidity.

Table 5-1 Relative Humidity Table

Dry bulb temperatures	Wet bulb temperatures															
	45	46	47	48	49	50	51	52	53	54	55	56	57	58	59	60
70	−14	0	+9	16	21	26	30	33	36	39	42	45	47	49	51	53
	3	6	9	12	16	19	22	26	29	33	36	40	44	48	51	55
71	−26	−5	+5	13	19	24	28	32	35	38	41	44	46	48	51	53
	2	5	8	11	14	17	20	23	27	30	34	37	41	45	48	52
72		−13	+1	10	16	22	26	30	34	37	40	43	45	47	50	52
		3	6	9	12	15	18	21	24	29	31	35	38	42	45	49
73		−26	−5	+6	13	19	24	29	32	36	39	41	44	47	49	51
		2	4	7	10	13	16	19	22	25	29	32	35	39	42	46
74			−13	+1	10	17	22	29	31	34	37	40	43	46	48	50
			3	6	8	11	14	17	20	23	26	30	33	36	40	43
75			−25	−4	+6	14	20	25	29	33	36	39	42	45	47	49
			2	4	7	10	12	15	18	21	24	27	31	34	37	40
76				−57	−12	+2	11	17	23	27	31	35	38	41	44	46	49
					3	5	8	11	14	16	19	22	25	28	31	35	38
77					−23	−3	+7	15	21	25	30	38	37	40	43	45	48
					2	4	7	9	12	15	18	20	23	26	29	32	35

Pressure 30″

Elevation 0–500

Here's another fun weather project. A simple wet bulb thermometer can be made from a regular thermometer, a bootlace, and a small medicine bottle. It's best to boil the bootlace before using it to get all the impurities and coloring matter out of it. Use about three inches of a tubular white cotton bootlace. Slip one end of it over the bulb of the thermometer. Part of the thermometer's wooden or metal backing may have to be cut off. Arrange the thermometer so that the other end of the bootlace dips into a small bottle filled with water. The water will soak up into the bootlace and keep the bulb moist. The wet bulb thermometer may then be mounted on a suitable backing with another thermometer to form a psychrometer. Make sure the temperature reading on the two thermometers is the same before they are purchased. A factory-built psychrometer (FIG. 5-8), built to Weather Service specifications, can be purchased for around $35.

Hygrometers should be kept out-of-doors in a properly ventilated shelter. In below-freezing temperatures, care should be taken outside between readings.

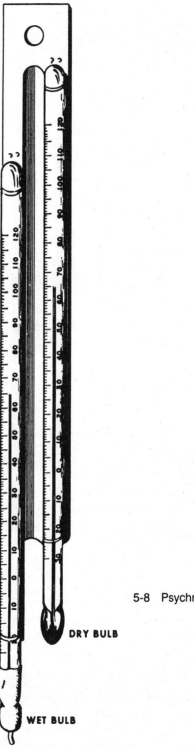

5-8 Psychrometer.

Sling Psychrometer

The *sling psychrometer* (FIG. 5-9) consists of a wooden grip with a swivel head and harness-type snap or spring clip for attaching to the top hole of the psychrometer frame.

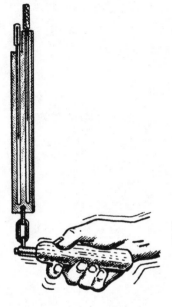

5-9 The sling psychrometer is swung in a circular motion before the reading.

When not in use, the sling psychrometer should be hung on a hook. Handle the sling psychrometer carefully at all times. The thermometers are easily broken through careless handling, dropping, or striking some object while being whirled.

For a check on humidity, take the instrument to a clear and open place exposed to the wind. Never touch the bulb or stem in handling or expose it to the direct rays of the sun while making an observation. The bulb of the wet bulb thermometer, which is covered with a wick, is moistened with clean water at the time an observation is made. Stand in a clear shady place facing into the wind and hold the psychrometer as far in front of the body as possible. Rotate the psychrometer with the wrist. Bring the psychrometer to a stop without any sharp jar and bring to eye level. Then read both thermometers to the nearest tenth of a degree, reading the wet bulb thermometer first. The whirling is repeated and other readings are make until two successive wet bulb readings are the same.

Rotor Psychrometer

The *rotor psychrometer* shown in FIG. 5-10 is a psychrometer attached to a handcrank-operated shaft. The rotor is designed for wall mounting and allows it to be permanently installed on the side wall of the instrument shelter. The same procedure is followed to get a correct reading of the rotor-mounted psychrome

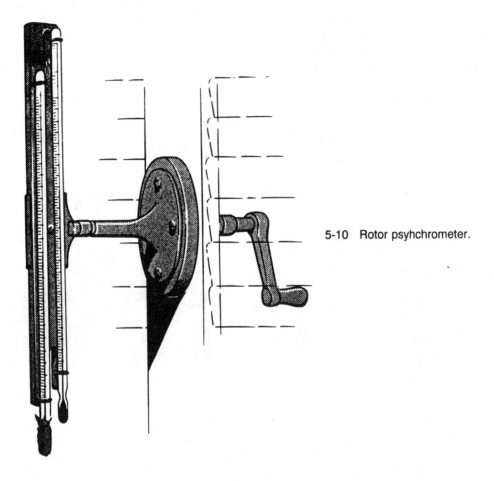

5-10 Rotor psyhchrometer.

ter as was given for the sling psychrometer, except that the instrument is left in the shelter. Hand electronic psychrometers are also used by both amateur and professional meteorologists (FIG. 5-11).

RAIN GAUGES

You can also make your own rain gauge for your weather station. Any straight-sided waterproof container can be used. The rainfall for any given period of time is the depth of the rain falling on a horizontal surface during that period. If the container has straight sides and the same area of cross section as the container opening, the depth of catch measured in inches and tenths is the correct rainfall at that location.

A common #10 can (6 1/16 inches inside diameter by 7 inches tall) will do very well for a low-cost rain gauge. (Canned fruits and vegetables often have the can size printed on the label.) The can should be located in an unsheltered place, keeping it level and protected from upsetting in the wind.

The depth of the water is then measured with a wooden ruler marked in inches and tenths. If a more accurate measurement is desired, the water can be

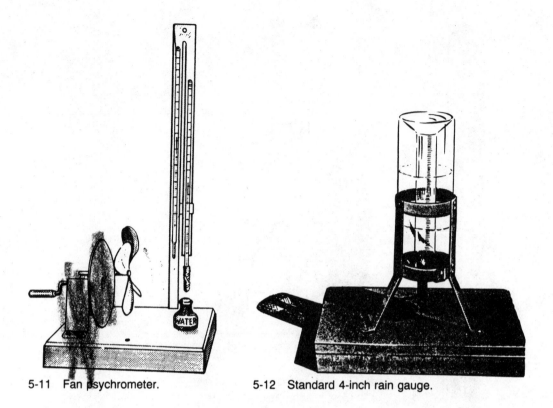

5-11   Fan psychrometer.   5-12   Standard 4-inch rain gauge.

poured from the #10 can into a #303 can, which happens to have an area of cross section about one-forth that of the larger can. The value found by measuring the precipitation when poured into the smaller can and divided by four will be more accurate than if measured directly in the large can.

An exposure site should be chosen which is a few feet from the ground and well away from buildings and other obstructions. Care should be taken to see that the rain does not splash into the can from the support post or any other nearby object.

The non-recording plastic rain gauge used by many amateur observation stations is made of a 4-inch diameter collector ring and funnel which fit over the top of a tripod mounted overflow container and a clear plastic measuring tube (FIG. 5-12). The rainfall drains from the collector ring through the funnel into the measuring tube, which gas graduations in the wall for direct reading to the nearest hundredth of an inch. The only maintenance required for the standard 4-inch rain gauge is to keep it clean and make sure it is mounted firmly.

MEASURING CLOUDS

Clouds, cloud ceilings, and visibility are important to the weather forecaster and are especially important to those who depend on visibility, such as aircraft pilots. Some of these observations require advanced instruments found in a professional weather station. But simple cloud observations can also be made with the human eye and mind.

First, a few terms should be defined. Clouds are a visible collection of particles of water, ice, or both in the free air. Clouds may also contain some foreign particles such as dust or smoke. Layer or cloud layer is the equal height of a large group of clouds, such as a cloud layer at 4000 feet. An *obstruction* means the observer cannot see more than 10 percent of the sky. Partial obstruction then must mean that some of the sky can be seen, measured in tenths. The *ceiling* is the height of the lowest opaque layer of clouds. A *low ceiling* means that clouds are low. An *unlimited ceiling* means that there are no opaque clouds. Finally, *visibility* is defined as the greatest distance at which selected objects can be seen and identified.

Visibility is decided by measuring the distance from the observation point to prominent landmarks at varied distances during a clear day. Then, when the fog comes rolling in you can accurately record that "visibility is one mile" by seeing the church steeple a mile away, but none of the farther objects (FIG. 5.13).

The simplest ceiling measurement devise is the *clinometer*, a portable hand instrument used to measure the height of a light "spot" projected onto the base of a cloud. A projector is mounted on the ground to project a light beam on to the bottom of a cloud above it. Then the clinometer is used to measure the distance from the spot to the instrument. Using the baseline, the height of the cloud base can then be figured.

Visibility is measured by an instrument called the *transmissometer*. It resembles an electric eye that reads out in variable figures to illustrate the degree of visibility.

When estimating visibility with the human eye and pre-selected landmarks you can rate the visibility as shown in TABLE 5.2. Simple visibility measurements such as this can be made two or three times a day, such as 8 a.m., 12 noon, and 4 p.m.

With the human eye and a few inexpensive or do-it-yourself weather instruments you can make accurate and useful observations of the atmosphere around you, the first important step to forecasting tomorrow's weather.

DO YOU KNOW...

- How to build your own weather shack?
- How wind direction is measured?
- How to build and use a weather vane?
- Why a barometer is important to weather forecasting?
- The difference between a *mercury* and *aneroid* barometer?
- How to build your own rain gauge?

92 BASIC INSTRUMENTS

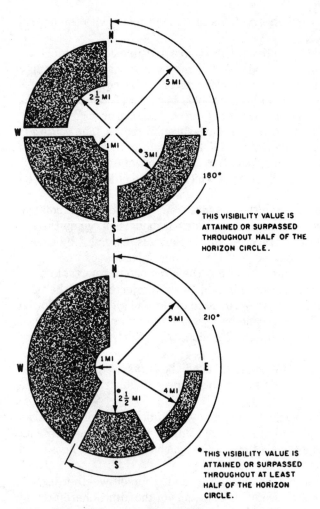

5-13 Determining prevailing visibility.

Table 5-2 Estimating Visibility

Distance	Visibility
200 yards	fog
1000 yards	mist or haze
1 mile	poor
5 miles	moderate
over 5 miles	good

6

Collecting Weather Data

THE FIRST STEP IN ACCURATELY FORECASTING WEATHER IS GATHERING ALL THE METEOROlogical data available on the area. The sources for such information include the weather eye and basic instruments discussed in Chapter 5. Before the data can be analyzed and interpreted it must be recorded.

THE WEATHER LOG

Don't trust your memory when making a weather observation: write down all observations. Try to take at least two readings of sky conditions and instruments each day. Additional readings will give an even better picture of changing weather patterns. Observations every 3 or 6 hours are acceptable for most amateur records. Keep a weather log to record your observations (FIG. 6-1). It can be as simple or as detailed as you want to make it, depending on the observations to be made.

MONTH	SKY		TEMPERATURE					HUMIDITY		BAROMETER		WIND		Precipitation		Remarks
	AM	PM	AM	PM	Max.	Min.	Mean	AM	PM	AM	PM	AM	PM	AM	PM	
1																
2																
3																
4																

6-1 Typical weather observation log.

COLLECTING WEATHER DATA

SKY CONDITION	PRESENT WEATHER		PRESSURE TENDENCY	CLOUDS
○ CLEAR	● RAIN	▽̇ RAIN SHOWER	∕‾ RISING, THEN FALLING	─ St
① 1/10 OR LESS	⁹ DRIZZLE		∕ RISING AND STEADY	◡ Sc
◔ 2/10 TO 3/10	✻ SNOW	▽ SQUALL	∕ RISING	∠ Ns
◔ 4/10	△ ICE PELLETS	)(FUNNEL CLOUD	✓ FALLING, THEN RISING	⌒ Cu
◐ 5/10	▽ HAIL	─+ BLOWING SNOW	─ STEADY	⌂ Cb
◕ 6/10	⚡ THUNDERSTORM	≡ FOG	_ FALLING, THEN RISING	◡ Ac
◕ 7/10 TO 8/10	⁹⁹ FREEZING DRIZZLE	↯ BLOWING DUST OR SAND	\ FALLING, THEN STEADY	∠ As (THIN)
● 9/10	⁹⁹ FREEZING RAIN	Ƨ DUST DEVIL	\ FALLING	⌒ Ci
● COMPLETE OVERCAST	✻̇ SNOW SHOWER	∿ SMOKE	\‾ RISING, THEN FALLING	⌇ Cc
⊗ OBSCURATION	⚡̇ THUNDERSTORM AND RAIN	∞ HAZE		⌇ Cs

6-2 The most common weather map symbols for sky condition, present weather, pressure tendency and clouds.

Here is a brief explanation of the entries to be made in a typical weather log using the basic instruments discussed in Chapter 5.

Sky: Enter the state of the sky (cloud cover) in tenths or substitute the following generally recognized weather symbols (FIG. 6-2):

- 0   Clear (less than 1/10).
- ◐   Scattered (2/10 to 5/10).
- ◑   Broken (5/10 to 9/10).
- ●   Overcast (more than 9/10).

If you see some form of precipitation, substitute one of the following letters for the sky symbol:

- R   Rain.
- S   Snow.
- H   Hail.
- T   Thunderstorm.
- E   Sleet.

- F   Fog.
- Z   Freezing rain.
- L   Drizzle.

Temperature: Read all thermometers to the nearest whole degree (°C or °F) and enter the current reading. Read the maximum thermometer at the P.M. observation and the minimum thermometer at the A.M. observation. Remember to reset these thermometers after each reading. Calculate the day's mean (average) temperature by adding the maximum and minimum temperatures and dividing by 2.

Humidity: Read the wet and dry bulb thermometers of the psychrometer, consult the Table of Relative Humidity (TABLE 6-1), and enter as a whole percentage. Compare the reading on the wet bulb thermometer with that on the dry bulb thermometer to find relative humidity on this table. Top figures on the chart are the present dry bulb reading. By checking the left column for degree difference shown on the two thermometers and across the dry bulb reading, if the difference between the dry and wet bulb readings is 6 degrees, and the dry bulb reading is 70 degrees, the relative humidity is 72 percent.

Barometer: Read the barometer to the nearest hundredth of an inch or half-millibar. If the barometer has been rising during the past three hours, mark a plus (+) after the reading entered; if falling, mark a minus (−) after the reading.

Wind: Check the wind direction (FIG. 6-3) to eight points of the compass (N, NE, E, SE, S, SW, W, NW). Read the anemometer or estimate the wind speed

Table 6-1 Table of Relative Humidity

Difference between wet-bulb and dry-bulb readings	Table of Relative Humidity — Temperature of air, dry-bulb thermometer, Fahrenheit							
	30°	40°	50°	60°	70°	80°	90°	100°
1	90	92	93	94	95	96	96	97
2	79	84	87	89	90	92	92	93
3	68	76	80	84	86	87	88	90
4	58	68	74	78	81	83	85	86
6	38	52	61	68	72	75	78	80
8	18	37	49	58	64	68	71	71
10		22	37	48	55	61	65	68
12		8	26	39	48	54	59	62
14			16	30	40	47	53	57
16			5	21	33	41	47	51
18				13	26	35	41	47
20				5	19	29	36	42
22					12	23	32	37
24					6	18	26	33

6-3 The wind rose.

(TABLE 6-2). Enter the wind direction and speed in the appropriate column, i.e., NE-5 or SW-12.

Precipitation: Read the gauge and enter the amount of precipitation to the nearest tenth of an inch. If no precipitation has occurred, leave the column blank. Show a reading of less than a tenth of an inch of precipitation by a "T" for "trace." If precipitation is snow, take several readings on the ground with a yardstick. Make sure the measurements are not made in drifted snow. Average the readings and enter the number of inches followed by the symbol "*" to mean snow. Generally speaking, ten inches of snow equals one inch of water or rain. Of course this varies, depending on the composition of the snow. Professional meteorologists often melt the snow to learn its moisture content. Be sure to empty the gauge after each observation.

Remarks: Enter any special phenomena such as lightning, thunder, smoke, or haze and note the times that the event was seen. Entries can also be made here for the type of clouds in the sky and the direction of movement. A common system of noting clouds is:

- CI   Cirrus.
- CC   Cirrocumulus.
- CS   Cirrostratus.
- AC   Altocumulus.

Table 6-2 Wind Speed Table

Wind speed			Seaman's term	World meteorological organization (1964)	Estimating wind speed			World meteorological organization		
Knots	Km per hour				Effects observed at sea	Effects observed on land		Term	Code	Height of waves in feet
under 1	under 1		Calm	Calm	Sea like mirror.	Calm; smoke rises vertically.		Calm, glassy	0	0
1–3	1–5		Light air	Light air	Ripples with appearance of scales; no foam crests.	Smoke drift indicates wind direction; vanes do not move.				
4–6	6–11		Light breeze	Light breeze	Small wavelets; crests of glassy appearance, not breaking.	Wind felt on face; leaves rustle; vanes begin to move.		Calm, rippled	1	0–⅓
7–10;	12–19		Gentle breeze	Gentle breeze	Large wavelets; crests begin to break; scattered whitecaps.	Leaves, small twigs in constant motion; light flags extended		Smooth, wavelets	2	⅓–1⅔
11–16	20–28		Moderate breeze	Moderate breeze	Small waves, becoming longer; numerous whitecaps.	Dust, leaves, and loose paper raised up; small branches move.		Slight	3	2–4
17–21	29–38		Fresh breeze	Fresh breeze	Moderate waves, taking longer form; many whitecaps; some spray.	Small trees in leaf begin in sway.		Moderate	4	4–8
22–27	39–49		Strong breeze	Strong breeze	Larger waves forming; whitecaps everywhere; more spray	Larger branches of trees in motion; whistling heard in wires.		Rough	5	8–13
28–33	50–61		Moderate gale	Near gale	Sea heaps up; white foam from breaking waves begins to be blown in streaks.	Whole trees in motion; resistance felt in walking against wind.		Very rough	6	13–20

Table 6-2 Wind Speed Table

Wind speed		Seaman's term	World meteorological organization (1964)	Estimating wind speed		World meteorological organization		
Knots	Km per hour			Effects observed at sea	Effects observed on land	Term	Code	Height of waves in feet
34–40	62–74	Fresh gale	Gale	Moderately high waves of greater length; edges of crests begin to break into spindrift; foam is blown in well-marked streaks.	Twigs and small branches broken off trees; progress generally impeded.			
41–47	75–88	Strong gale	Strong gale	High waves; sea begins to roll; dense streaks of foam; spray may reduce visibility.	Slight structural damage occurs; slate blown from roofs.			
48–55	89–102	Whole gale	Storm	Very high waves with overhanging crests; sea takes white appearance as foam is blown in very dense streaks; rolling is heavy and visibility reduced.	Seldom experienced on land; trees broken or uprooted; considerable structural damage occurs.	High	7	20–30
56–63	103–117	Storm	Violet storm	Exceptionally high waves; sea covered with white foam patches; visibility still more reduced.		Very high	8	30–45
64–71 72–80 81–89 90–99 100–108 109–118	118–133 134–149 150–166 167–183 184–201 202–220	Hurricane	Hurricane	Air filled with foam; sea completely white with driving spray; visibility greatly reduced.	Very rarely experienced on land; usually accompanied by widespread damage.	Phenomenal	9	over 45

- AS   Altostratus.
- NS   Nimbostratus.
- SC   Stratovumulus.
- ST   Stratus.
- CU   Cumulus.
- CB   Cumulonimbus.

You may want to use the abbreviations and International Code signs used by the National Weather Service.

COLLECTING DATA FROM THE WEATHER SERVICE

The amateur meteorologist can expand weather observations with information from the world's largest system of weather data collection, the National Weather Service. Forecasts, outlooks, and raw data are available from the Weather Service in a number of ways.

Forecasts

Forecasts are detailed predictions of expected weather conditions for periods of time up to an individual city, Chicago forecast, or an area as large as a state, Vermont forecast.

Local, zone, and state forecasts are made several times a day to predict the weather, in detail, for the next 36 to 48 hours. A *local* forecast applies to a city and its immediate surroundings. A *zone* forecast applies to a small portion of a state. A *state* forecast applies to an entire state (TABLE 6-3). Extended forecasts are issued daily to predict weather over a state in broad terms for a period three days beyond that of the state forecast.

Travelers' forecasts are written in two forms. One is issued nationwide for about 90 cities and consists of a one word statement on the expected weather and the forecast high and low temperature for the next 36 to 48 hours for each city. The other is issued by some local National Weather Service offices to describe general driving conditions over routes within a few hundred miles of the city where the office is located.

Recreation forecasts are issued by selected NWS offices based on local needs and office capabilities. Forecasts for boating and skiing areas are two of the more popular examples.

Weather watches and warnings are issued in connection with hazardous weather: tornadoes, winter storms, hurricanes, flash floods, and the like. A *watch* is issued, where possible, well in advance to alert the public to the possibility of dangerous weather developing. A *warning* is normally issued when the hazard is near.

River and flood forecasts and warnings provide daily information on stages of major rivers and streams and, when necessary, warnings of potential flood conditions.

Table 6-3 Time Conversion Table

PREVIOUS DAY							SAME DAY																		NEXT DAY												
1800	1900	2000	2100	2200	2300	2400	0100	0200	0300	0400	0500	0600	0700	0800	0900	1000	1100	1200	1300	1400	1500	1600	1700	1800
1900	2000	2100	2200	2300	2400	0100	0200	0300	0400	0500	0600	0700	0800	0900	1000	1100	1200	1300	1400	1500	1600	1700	1800	1900
2000	2100	2200	2300	2400	0100	0200	0300	0400	0500	0600	0700	0800	0900	1000	1100	1200	1300	1400	1500	1600	1700	1800	1900	2000
2100	2200	2300	2400	0100	0200	0300	0400	0500	0600	0700	0800	0900	1000	1100	1200	1300	1400	1500	1600	1700	1800	1900	2000	2100
2200	2300	2400	0100	0200	0300	0400	0500	0600	0700	0800	0900	1000	1100	1200	1300	1400	1500	1600	1700	1800	1900	2000	2100	2200
2300	2400	0100	0200	0300	0400	0500	0600	0700	0800	0900	1000	1100	1200	1300	1400	1500	1600	1700	1800	1900	2000	2100	2200	2300
2400	0100	0200	0300	0400	0500	0600	0700	0800	0900	1000	1100	1200	1300	1400	1500	1600	1700	1800	1900	2000	2100	2200	2300	2400
0100	0200	0300	0400	0500	0600	0700	0800	0900	1000	1100	1200	1300	1400	1500	1600	1700	1800	1900	2000	2100	2200	2300	2400	0100
0200	0300	0400	0500	0600	0700	0800	0900	1000	1100	1200	1300	1400	1500	1600	1700	1800	1900	2000	2100	2200	2300	2400	0100	0200
0300	0400	0500	0600	0700	0800	0900	1000	1100	1200	1300	1400	1500	1600	1700	1800	1900	2000	2100	2200	2300	2400	0100	0200	0300
0400	0500	0600	0700	0800	0900	1000	1100	1200	1300	1400	1500	1600	1700	1800	1900	2000	2100	2200	2300	2400	0100	0200	0300	0400
0500	0600	0700	0800	0900	1000	1100	1200	1300	1400	1500	1600	1700	1800	1900	2000	2100	2200	2300	2400	0100	0200	0300	0400	0500
0600	0700	0800	0900	1000	1100	1200	1300	1400	1500	1600	1700	1800	1900	2000	2100	2200	2300	2400	0100	0200	0300	0400	0500	0600
0700	0800	0900	1000	1100	1200	1300	1400	1500	1600	1700	1800	1900	2000	2100	2200	2300	2400	0100	0200	0300	0400	0500	0600	0700
0800	0900	1000	1100	1200	1300	1400	1500	1600	1700	1800	1900	2000	2100	2200	2300	2400	0100	0200	0300	0400	0500	0600	0700	0800
0900	1000	1100	1200	1300	1400	1500	1600	1700	1800	1900	2000	2100	2200	2300	2400	0100	0200	0300	0400	0500	0600	0700	0800	0900
1000	1100	1200	1300	1400	1500	1600	1700	1800	1900	2000	2100	2200	2300	2400	0100	0200	0300	0400	0500	0600	0700	0800	0900	1000
1100	1200	1300	1400	1500	1600	1700	1800	1900	2000	2100	2200	2300	2400	0100	0200	0300	0400	0500	0600	0700	0800	0900	1000	1100
1200	1300	1400	1500	1600	1700	1800	1900	2000	2100	2200	2300	2400	0100	0200	0300	0400	0500	0600	0700	0800	0900	1000	1100	1200
1300	1400	1500	1600	1700	1800	1900	2000	2100	2200	2300	2400	0100	0200	0300	0400	0500	0600	0700	0800	0900	1000	1100	1200	1300
1400	1500	1600	1700	1800	1900	2000	2100	2200	2300	2400	0100	0200	0300	0400	0500	0600	0700	0800	0900	1000	1100	1200	1300	1400
1500	1600	1700	1800	1900	2000	2100	2200	2300	2400	0100	0200	0300	0400	0500	0600	0700	0800	0900	1000	1100	1200	1300	1400	1500
1600	1700	1800	1900	2000	2100	2200	2300	2400	0100	0200	0300	0400	0500	0600	0700	0800	0900	1000	1100	1200	1300	1400	1500	1600
1700	1800	1900	2000	2100	2200	2300	2400	0100	0200	0300	0400	0500	0600	0700	0800	0900	1000	1100	1200	1300	1400	1500	1600	1700
Y	X	W	V	U	T	S	R	Q	P	O	N	Z	A	B	C	D	E	F	G	H	I	K	L	M
+12	+11	+10	+9	+8	+7	+6	+5	+4	+3	+2	+1	0	−1	−2	−3	−4	−5	−6	−7	−8	−9	−10	−11	−12

Outlooks

Outlooks are estimates of general weather conditions for periods beyond five days from the date of issue. Naturally, such predictions have less precision and detail than short-term forecasts. They are useful for general planning.

Six to ten day outlooks are issued three times a week. They give estimates of temperature an precipitation for each state in very broad terms with respect to normal.

Average monthly weather outlooks are issued shortly before the 1st and 15th of each month. They deal with changes from the average in temperature and precipitation over broad areas of the nation for the next 30 days.

Seasonal outlooks are issued during the last week of February, May, August, and November, projecting temperature changes from the average for the following three months. They occasionally include a precipitation outlook.

Raw Weather Data.

National Weather Service offices can also be your source of weather data. Through various distribution methods the NWS offers current information on precipitation, temperature, pressure, wind direction and speed, humidity, dew point, clouds, visibility, ceiling, and other valuable data. In fact, by using weather information distributed free or at minimal cost, you can forecast local and even regional weather accurately without a single weather observation instrument.

DISTRIBUTION OF WEATHER INFORMATION

National Weather Service products are distributed in a number of ways. In addition to the systems operated by local weather offices, information is given to the news media for relay to the public.

The primary outlet for NWS forecasts and warnings is radio, television, and newspapers. Coverage varies greatly from station to station and paper to paper because reporting weather information is a voluntary service. Many cable TV systems have a channel devoted entirely or in part to weather. Figure 6-4 shows a typical newspaper's weather column. It offers area weather conditions, river and tide conditions, and a simplified national weather map.

Printed weather information is also available directly from the National weather Service. The Average Monthly Weather Outlook, including the Seasonal Outlook, Daily Weather Maps, Weekly Series, and other publications are available through the Superintendent of Documents, U.S. Government Printing Office, Washington, D.C., 20402, on a fee basis. Subject Bibliography (SB) 234, entitled *Weather*, is available from the same source. It offers a list and prices of meteorological publications available through the government.

Telephone recordings are available at many National Weather Service offices. For a listing of these numbers see your city's phone directory under "United States Government, Department of Commerce, National Weather Ser-

WEATHER

In our area

TEMPERATURE	High	Low
SW Research Center, Hazel Dell		
In 24 hours to 8 a.m.	96	53
Weather Service, Portland		
In 24 hours to 8 a.m.	99	59
Yesterday's record	95	48
	1958	1975

PRECIPITATION	Inches
SW Research Center, Hazel Dell	
In 24 hours to 8 a.m.	none
Cumulative this year	24.45
To this date last year	20.12
Weather Service, Portland	
In 24 hours to 8 a.m.	none

River level

The Columbia River reached 13.1 feet below flood stage at 7 a.m. today at the Interstate 5 Bridge.

Local skies

WEDNESDAY, AUGUST 25

- Sunset today . . . . 8:03 p.m.
- Sunrise tomorrow . . . . 6:24 a.m.
- Moonset tonight . . . . 11:40 p.m.
- First quarter tomorrow . . . . . 2:49 a.m.

The planet Mercury is low in the west at sunset and sets in the early evening twilight. Do not be disappointed if you cannot see the elusive planet; it is making about its poorest appearance of the year.

Forecasts

Vancouver — Portland — Fair and cooler today with highs in the upper 80s. Fair Thursday after morning clouds. Highs near 80. Low 60. Southwest to west winds 10-20 mph.

Vancouver north to Olympia — Fair tonight except for patchy fog or low clouds, lows in the 50s and winds variable 5-15 mph. Fair Thursday after morning fog or low clouds, highs in the 70s.

Willamette Valley — Partly cloudy tonight. Lows 55 to 60. Variable cloudiness. Highs about 80.

Columbia Gorge — Variable cloudiness through Thursday. Lows near 60. Highs low 80s west to low 90s east. West wind to 20 mph.

Regional

24 hours to 4 a.m. today

	High	Low	Pr
Astoria	82	58	
Bellingham	80	56	
Colville	89	57	
Eugene	96	54	
Hoquiam	90	58	
Lakeview	92	53	
Medford	101	59	
Newport	62	52	
North Bend	64	57	
Olympia	94	52	
Portland	99	63	
Redmond	90	53	
Salem	97	54	
Seattle	88	62	
Spokane	90	55	
Walla Walla	96	64	
Wenatchee	93	62	
Yakima	90	51	

National

24 hours to 4 a.m. today PDT

	High	Low	Prc	Otlk
Albany	82	58	.01	clr
Albuquerque	93	64	.16	cdy
Amarillo	88	59	.20	cdy
Anchorage	61	49	.07	cdy
Asheville	84	67	.51	cdy
Atlanta	92	73		cdy
Atlantic City	80	68	.51	clr
Austin	100	73		clr
Baltimore	87	71	.12	clr
Billings	82	60		cdy
Birmingham	91	72		clr
Bismarck	74	49		cdy
Boise	66	55		cdy
Boston	85	66		cdy
Brownsville	96	79		cdy
Buffalo	76	62	1.46	cdy
Burlington	77	61		clr
Casper	78	50		cdy
Charleston, S.C.	94	80		cdy
Charleston, W.Va.	85	72	.07	clr
Charlotte, N.C.	93	75	.26	clr
Cheyenne	69	49		clr
Chicago	75	54	1.06	cdy
Cincinnati	82	68	.17	clr
Cleveland	84	65	.13	clr
Columbia, S.C.	94	75		cdy
Columbus	83	69	.16	clr
Dallas-Fort Worth	101	76		clr
Dayton	81	65	1.54	clr
Denver	74	55		cdy
Des Moines	79	55	.03	cdy
Detroit	80	61	.01	cdy
Duluth	72	43	.09	cdy
El Paso	96	71	.05	cdy
Fairbanks	73	49		cdy
Fargo	74	49		cdy
Flagstaff	77	56	.39	rn
Great Falls	78	56	1.26	cdy
Hartford	84	62	.02	clr
Helena	86	48		cdy
Honolulu	89	75		clr
Houston	94	78		clr
Indianapolis	78	63	.25	cdy
Jackson, Miss.	95	74		clr
Jacksonville	95	75	.15	cdy
Juneau	58	51	.20	rn
Kansas City	79	56	.21	cdy
Knoxville	86	77	.10	cdy
Las Vegas	80	68	.22	cdy
Little Rock	96	75		cdy
Los Angeles	75	66		cdy
Louisville	86	68	.04	cdy
Lubbock	97	65		cdy
Memphis	94	80		rn
Miami	88	83		cdy
Milwaukee	70	55	.35	cdy
Minneapolis-St. Paul	76	53	.56	cdy
Nashville	90	71	.70	rn
New Orleans	95	73		cdy
New York	84	70	.30	clr
Norfolk	85	74		clr
North Platte	75	47		cdy
Oklahoma City	102	69		cdy
Omaha	74	54	.18	cdy
Orlando	93	76		clr
Philadelphia	85	70	.19	clr
Phoenix	96	75	.65	cdy
Pittsburgh	81	68	.52	clr
Portland, Maine	80	60	.01	clr
Providence	83	63		clr
Raleigh	91	75		cdy
Rapid City	73	50		cdy
Reno	97	54		clr
Richmond	87	73		clr
Salt Lake City	89	65		clr
San Antonio	100	76		clr
San Diego	74	71		cdy
San Francisco	60	54		clr
Shreveport	96	74		cdy
Sioux Falls	76	53		cdy
St. Louis	82	59	.05	cdy
St. Pete-Tampa Bay	91	78		cdy
Sault Ste. Marie	63	42		cdy
Syracuse	80	61	.35	cdy
Topeka	80	53	.09	cdy
Tucson	88	66	.35	cdy
Tulsa	104	69		cdy
Washington	87	74		clr
Wichita	86	62		cdy

Extremes, excluding Alaska:
High: 105   Fort Sill, Oklahoma
Low: 30   West Yellowstone, Montana

Precipitation — Pr.   Rain — rn
Outlook — Otlk.   Snow — sn
Cloudy — cdy   Clear — clr

Tides

High Tides at Astoria

Aug. 26	6:59 a.m.,	5.5	6:55 p.m.,	7.2
Aug. 27	8:09 a.m.,	5.3	7:53 p.m.,	7.1
Aug. 28	9:21 a.m.,	5.4	8:52 p.m.,	7.1
Aug. 29	10:23 a.m.,	5.6	9:51 p.m.,	7.2
Aug. 30	11:12 a.m.,	5.9	10:43 p.m.,	7.4
Aug. 31	11:58 a.m.,	6.3	11:30 p.m.,	7.6
Sept 1			12:36 p.m.,	6.6

Low Tides at Astoria

Aug. 26	0:54 a.m.,	0.7	12:39 p.m.,	2.2
Aug. 27	1:53 a.m.,	0.6	1:38 p.m.,	2.7
Aug. 28	2:55 a.m.,	0.5	2:45 p.m.,	2.9
Aug. 29	3:56 a.m.,	0.2	3:51 p.m.,	2.9
Aug. 30	4:51 a.m.,	-0.1	4:48 p.m.,	2.7
Aug. 31	5:38 a.m.,	-0.4	5:41 p.m.,	2.4
Sept 1	6:19 a.m.,	-0.6	6:25 p.m.,	2.0

Deduct these hours and minutes from tables: Long Beach, 1:10; Clatsop beaches, 0:50; Tillamook, 0:45; Yaquina, 1:00; Newport, 1:15; Vancouver add 5:45 for high, 7:24 for low.

The Forecast For 8 p.m. EDT Thursday, August 26
● High Temperatures
Rain, Snow, Showers, Flurries

National Weather Service
NOAA, U.S. Dept. of Commerce

Fronts: Cold ▼▼   Warm ●●   Occluded ▼●   Stationary ●■

6-4 Typical newspaper weather map and information.

vice," or simply "Weather." The long-distance information operator can provide you with the numbers of weather recordings in other cities, where available.

Weather Radio

The National Oceanic and Atmospheric Administration (NOAA), parent of the National Weather Service, offers continuous weather information directly from about 350 NWS offices. Taped weather messages are revised every one to three hours, or more frequently if needed. Most of the stations operate 24 hours daily.

The broadcasts are tailored to the weather information needs of people within the receiving area. For example, stations along the sea coasts and Great Lakes provide specialized weather information for boaters and fishermen.

During severe weather, National Weather Service forecasters can interrupt the routine weather broadcasts and substitute special warning messages. The forecasters can also activate specially designed warning receivers. Such receivers either sound an alarm indicating that an emergency exists and alert the listener to turn the receiver up or, when operated in a muted mode, automatically turns it up so the message can be heard. "Warning Alarm" receivers are especially valuable for schools, hospitals, public safety agencies, and news media offices.

NOAA Weather Radio broadcasts are made on one of seven high-band FM frequencies ranging from 162.40 to 162.55 megahertz (MHz). These frequencies are not found on the average home radio, however, a number of radio manufacturers offer special weather radios that operate on these frequencies, with or without the emergency warning alarm. Also, there are now many radios on the market that offer standard AM/FM frequencies plus the so-called "weather band" as an added feature.

Table 6-4 offers a complete listing of stations in the NOAA Weather Radio Network. Figure 6-5 is a map illustrating locations of these stations. These broadcasts can usually be heard as far as 40 miles or more from the antenna site. The effective range depends on many factors including the height of the broadcasting antenna, terrain, quality of the receiver, and type of receiving antenna. As a general rule, listeners close to or perhaps beyond the 40 mile range should have a good quality receiver system if they expect reliable reception. An outside antenna may also be needed in fringe areas. If practicable, the receiver should be tried at its place of intended use before making a final purchase. About 90 percent of the nation's population is within listening range of a weather band broadcast.

Teletype

Many aviators use the teletype as their source of weather information. It can also be used by other citizens to collect weather data. The Federal Aviation Administration and the National Weather Service work together to supply and man aviation weather offices in airports across the nation. These offices are called Flight Service Stations, abbreviated FSS. These Flight Service Stations provide

Table 6-4 National Weather Radio Network Station Locations and Frequencies

Legend—Frequencies are identified as follows:
(1)—162,550 MHz
(2)—162,400 MHz
(3)—162,475 MHz
(4)—162,425 MHz
(5)—162,450 MHz
(6)—162,500 MHz
(7)—162,525 MHz

Location	Frequency
Alabama	
Anniston	3
Birmingham	1
*Columbia	4
Demopolls	3
Dozier	1
Florence	3
Huntsville	2
Louisville	3
Mobile	1
Montgomery	2
Tuscaloosa	2
Alaska	
Anchorage	1
Cordova	1
Fairbanks	1
Homer	2
Juneau	1
Ketchikan	1
Kodiak	1
Nome	1
Petersburg	1
Seward	1
Sitka	2
Valdez	1
Wrangell	2
Yukutat	1
Arizona	
Flagstaff	2
Phoenix	1
Tucson	2
Yuma	1
Arkansas	
Mountain View	2
Fayetteville	3
Fort Smith	2
Gurdon	3
Jonesboro	1
Little Rock	1
Star City	2
Texarkana	1
California	
Bakersfield (P)	1
Coachella (P)	2
Eureka	2
Fresno	2
Los Angeles	1
Merced	1
Monterey	2
Point Arena	2
Redding (P)	1
Sacramento	2
San Diego	2
San Francisco	1
San Luis Obispo	1
Santa Barbara	2
Colorado	
Alamosa (P)	3
Colorado Springs	3
Denver	1
Grand Junction	1
Greeley	2
Longmont	1
Pueblo	2
Sterling	2
Connecticut	
Hartford	3
Meriden	2
New London	1
Delaware	
Lewes	1
District of Columbia	
Washington, D.C.	1
Florida	
Clewiston	2
Daytona Beach	2
Fort Myers	3
Gainesville	3
Jacksonville	1
Key West	2
Melbourne	1
Miami	1
Orlando	3
Panama City	1
Pensacola	2
Tallahassee	2
Tampa	1
West Palm Beach	3
Georgia	
Athens	2
Atlanta	1
Augusta	1
*Baxley	7
Chatsworth	2
Columbus	2
Macon	3
Pelham	1
Savannah	2
Waycross	3
Hawaii	
Hilo	1
Honolulu	1
Kokee	2
Mt. Haleakala	2
Walmanalo (R)	2
Idaho	
Boise	1
Lewiston (P)	1
Pocatello	1
Twin Falls	2
Illinois	
Champaign	1
Chicago	1
Marion	4
Moline	1
Peoria	3
Rockford	3
Springfield	2
Indiana	
Evansville	1
Fort Wayne	1
Indianapolis	1
Lafayette	3
South Bend	2
Terre Haute	2
Iowa	
Cedar Rapids	3
Des Moines	1
Dubuque (P)	2
Sioux City	3
Waterloo	1
Kansas	
Chanute	2
Colby	3
Concordia	1
Dodge City	3
Ellsworth	2
Topeka	3
Wichita	1
Kentucky	
Ashland	1
Bowling Green	2
Covington	1
Elizabethtown (R)	2
Hazard	3
Lexington	2
Louisville	3
Mayfield	3
Pikeville (R)	2
Somerset	1
Louisiana	
Alexandria	3
Baton Rouge	2
Buras	3
Lafayette	1
Lake Charles	2
Monroe	1
Morgan City	3
New Orleans	1
Shreveport	2
Maine	
*Dresden	3
Ellsworth	2
Portland	1
Maryland	
Baltimore	2
Hagerstown	3
Salisbury	3
Massachusetts	
Boston	3
Hyannis	1
Worcester	1
Michigan	
Alpena	1
Detroit	1
Flint	2
Grand Rapids	1
Houghton	2
Marquette	1
Onondaga	2
Sault Sainte Marie	1
Traverse City	2
Minnesota	
Detroit Lakes	3
Duluth	1
International Falls	1
Mankato	2
Minneapolis	1
Rochester	3
Saint Cloud (P)	3
Thief River Falls	1
Willmar (P)	2
Mississippi	
Ackerman	3
Boonesville	1
Bude	1
Columbia (R)	2
Gulfport	2
Hattiesburg	3
Inverness	1
Jackson	2
Meridian	1
Oxford	2
Missouri	
Columbia	2
Camdenton	1
Hannibal	3
Joplin/Carthage	1
Kansas City	1

Table 6-4 National Weather Radio Network Station Locations and Frequencies (Cont.)

Location	Freq
St Joseph	2
St Louis	1
Sikeston	2
Springfield	2
Montana	
Billings	1
Butte	1
Glasgow	1
Great Falls	1
Havre (P)	2
Helena	2
Kalispell	1
Miles City	2
Missoula	2
Nebraska	
Bassett	3
Grand Island	2
Holdrege	3
Lincoln	3
Merriman	2
Norfolk	1
North Platte	1
Omaha	2
Scottsbluff	1
Nevada	
Elko	1
Ely	2
Las Vegas	1
Reno	1
Winnemucca	2
New Hampshire	
Concord	2
New Jersey	
Atlantic City	2
New Mexico	
Albuquerque	2
Clovis	3
Des Moines	1
Farmington, 3	
Hobbs	2
Las Cruces	2
Ruidoso	1
Santa Fe	1
New York	
Albany	1
Binghamton	3
Buffalo	1
Elmira	1
Kingston	3
New York City	1
Rochester	2
Syracuse	1
North Carolina	
Asheville	2
Cape Hatteras	3
Charlotte	3
Fayetteville	3
New Bern	2
Raleigh/Durham	1
Rocky Mount	3
Wilmington	1
Winston-Salem	2
North Dakota	
Bismarck	2
Dickinson	2
Fargo	2
Jamestown	2
Minot	2
Petersburg	2
Williston	2
Ohio	
Akron	2
Caldwell	3
Cleveland	1
Columbus	1
Dayton	3
Lima	2
Sandusky	2
Toledo	1
Oklahoma	
Clinton	3
Enid	3
Lawton	1
McAlester	3
Oklahoma City	2
Tulsa	1
Oregon	
Astoria	2
Brookings	1
Coos Bay	2
Eugene	2
Klamath Falls	1
Medford	2
Newport	1
Pendleton	1
Portland	1
Roseburg	3
Salem	3
Pennsylvania	
Allentown	2
Clearfield	1
Erie	2
Harrisburg	1
Johnstown	2
Philadelphia	3
Pittsburgh	1
State College	3
Wilkes-Barre	1
Williamsport	2
Puerto Rico	
Maricao	1
San Juan	2
Rhode Island	
Providence	2
South Carolina	
Beaufort	3
Charleston	1
Columbia	2
Florence	1
Greenville	1
Myrtle Beach	2
Sumter (R)	3
South Dakota	
Aberdeen	3
Huron	1
Pierre	2
Rapid City	1
Sioux Falls	2
Tennessee	
Bristol	1
Chattanooga	1
Cookeville	2
Jackson	1
Knoxville	3
Memphis	3
Nashville	1
Shelbyville	3
Waverly	2
Texas	
Abilene	2
Amarillo	1
Austin	2
Beaumont	3
Big Spring	3
Brownsville	1
Bryan	1
Corpus Christi	1
Dallas	2
Del Rio (P)	2
El Paso	3
Fort Worth	1
Galveston	1
Houston	2
Laredo	3
Lubbock	2
Lufkin	1
Midland	2
Paris	1
Pharr	2
San Angelo	1
San Antonio	1
Sherman	3
Tyler	3
Victoria	2
Waco	3
Wichita Falls	3
Utah	
Logan	2
Cedar City	2
Vernal	2
Salt Lake City	1
Vermont	
Burlington	2
Windsor	3
Virginia	
Heathsville	2
Lynchburg	2
Norfolk	1
Richmond	3
Roanoke	3
Washington	
Neah Bay	1
Olympia	3
Seattle	1
Spokane	2
Wenatchee	3
Yakima	1
West Virginia	
(see note 5)	
Charleston	2
Clarksburg	1
Wisconsin	
La Crosse (P)	1
Green Bay	1
Madison	1
Menomonie	2
Milwaukee	2
Wausau	3
Wyoming	
Casper	1
Cheyenne	3
Lander	3
Sheridan (P)	3

Notes:
1. Stations marked with an asterisk (*) are funded by public utility companies.
2. Stations marked (R) are low powered experimental repeater stations serving a very limited local area.
3. Stations marked (P) operate less than 24 hours/day; however, hours are extended when possible during severe weather.
4. Occasionally the frequency of an existing or planned station must be changed because of unexpected radio frequency interference with adjacent NOAA Weather Radio stations and/or with other government or commercial operators within the area.
5. Six additional stations are planned for West Virginia, Gilbert, Beckley, Hinton, Spencer, Sutton and Romney. Frequencies had not been assigned at the time of this printing.

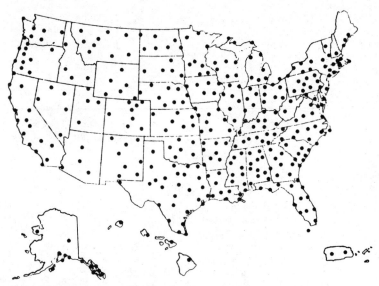

6-5 NOAA National Weather Radio Network station locations.

weather information to the public, but primarily to fliers, in person and over two-way radios.

Much of the FSS' weather information comes to them over teletypes from cities and stations across the nation. A Terminal Forecast offers information for specific airports on ceiling, cloud heights, cloud amounts, visibility, weather, and obstructions to vision, surface wind, and flight conditions. The terminal forecasts have a 24-hour valid period and are issued three times a day.

Area Forecasts are 18-hour forecasts plus a 12-hour outlook written twice a day with general details of cloud, weather, and frontal conditions for an are the size of several states.

READING THE WEATHER MAP

The most common and useful source of present and future weather conditions is sthe weather map, also called the *Surface Syntopic Weather Map*. These maps are available through the government on a subscription basis, in a simplified form in newspapers, or can be viewed at Flight Service Stations and National Weather Service offices. Figure 6-4 shows one of these weather maps.

Many surface weather maps made up each day by the National Meteorological Center (NMC). These maps are based on data recorded for an area. They provide a map-oriented picture of the weather as it exists at the time of observation. This picture includes the atmospheric pressure patterns at the surface, fronts, and individual reporting station data.

Facsimile syntopic maps are drawn at the NMC eight times each day. The observation time appears in the lower left corner of the map. The map is drawn to a 1:10,000,000 scale and shows a great number of stations over the North American area. They are then transmitted either by facsimile transmission equipment, fax machines, or by computer to stations around the United States.

Three or four sequence maps are displayed in the weather office to note trends in the development of fronts, pressure systems, and weather phenomena.

THE STATION MODEL

Data seen by individual stations are plotted using standard symbols on the surface weather map. These symbols are grouped around a printed circle called the *station circle*, which shows the geographical location of the station on the chart. This grouping is the *station model*. The arrangement of plotted data around the station model is standard in weather facsimile maps. The abbreviated station model includes only the sky cover, surface wind, present weather, temperature, dew point, cloud types, pressure, pressure tendency, and accumulated precipitation.

The direction from which the wind is blowing is shown on the surface weather map by a shaft. The shaft on the station models in FIG. 6-6 shows wind from the northwest. Wind speed is noted by the barb and pennant system: a full barb is valued at 10 knots, a half-barb at 5 knots, and a pennant at 50 knots. To avoid confusion in length with a single full barb, the half-barb appears part way down the shaft when 5-knot winds exist. The wind speed is a summation of the number of barbs and pennants appearing on the shaft. For a calm wind, a circle is drawn around the station center. The wind direction is from true north (a true wind), and the speed is rounded to the nearest 5 knots.

6-6 Facsimile station model.

The amount of black in the facsimile station circle is an indication of the total amount of sky coverage by all cloud layers over the station (FIG. 6-6). These symbols are part of the international surface syntopic code.

The mean sea level pressure is noted to the nearest millibar rounded off to two digits. If the reported pressure begins with a digit from 0 through 5, the figure is preceded with a 10. If the report pressure begins with a digit from 6 through 9, the figure is preceded by a 9, for example 99 means 999 millibars and 65 means 965 millibars. These pressure values are corrected to mean sea level to

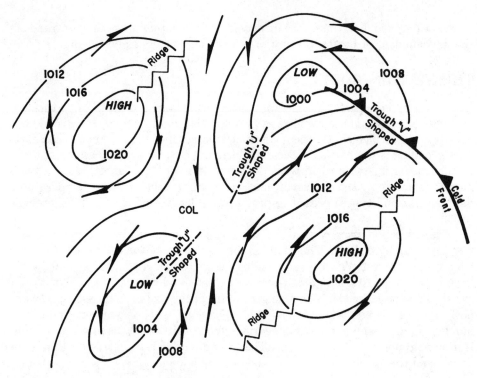

6-7   Basic isobaric patterns.

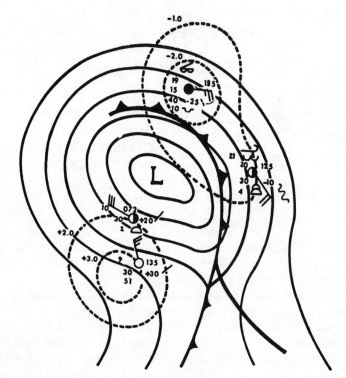

6-8   Isallobars.

eliminate pressure variations caused by different field elevations (FIGS. 6-7 and 6-8).

On surface weather maps prepared by the United States weather services the temperatures and dew points are given in degrees Fahrenheit (°F). However, weather maps produced in most other major countries are given in degrees Celsius (°C) on the centigrade scale.

The symbols used for present weather, illustrated in FIGS. 6-9 and 6-10, represent forms of precipitation or obstructions to vision, or both. Although there are 100 possible combinations and types of weather symbols, only a few are commonly used.

Pressure changes over the past three hours are also recorded. A plus or minus shows whether the net change in pressure has resulted in an increase or decrease. The amount of change is shown in tenths of millibars, and the pattern of change is shown by the coded symbol.

Although there are many cloud types symbolized in the international code, a knowledge of the ten basic types will be adequate. The same symbol, plotted above the station circle, is used for nimbostratus or thick altostratus clouds; the two cloud types are similar in that both may obscure the Sun and produce continuous precipitation at the surface.

Once collected and recorded, weather data is interpreted following the rules of meteorology. These forecast of future weather conditions can then be used to benefit pilots, farmers, business, students, parents, and citizens.

DO YOU KNOW...

- The reason for keeping a weather log?
- How to enter data in a weather log?
- The common way to write cloud information?
- The difference between a *weather watch* and a *weather warning*?
- The radio frequency of a nearby weather station?
- How to read a weather map?

ww
Present weather

00	01	02	03	04	05	06	07
Cloud development NOT observed or NOT observable during past hour. §	Clouds generally dissolving or becoming less developed during past hour. §	State of sky on the whole unchanged during past hour. §	Clouds generally forming or developing during past hour. §	Visibility reduced by smoke.	Dry haze.	Widespread dust in suspension in the air, NOT raised by wind, at time of observation.	Dust or sand raised by wind, at time of ob.
10	11	12	13	14	15	16	17
Light fog.	Patches of shallow fog at station, NOT deeper than 6 feet on land.	More or less continuous shallow fog at station, NOT deeper than 6 feet on land.	Lightning visible, no thunder heard.	Precipitation within sight, but NOT reaching the ground at station.	Precipitation within sight, reaching the ground, but distant from station.	Precipitation within sight, reaching the ground, near to but NOT at station.	Thunder heard, but no precipitation at the station.
20	21	22	23	24	25	26	27
Drizzle (NOT freezing and NOT falling as showers) during past hour, but NOT at time of ob.	Rain (NOT freezing and NOT falling as showers) during past hour, but NOT at time of ob.	Snow (NOT falling as showers) during past hr, but NOT at time of ob.	Rain and snow (NOT falling as showers) during past hour, but NOT at time of observation.	Freezing drizzle or freezing rain (NOT falling as showers) during past hour, but NOT at time of observation.	Showers of rain during past hour, but NOT at time of observation.	Showers of snow, or of rain and snow, during past hour, but NOT at time of observation.	Showers of hail, or of hail and rain, during past hour, but NOT at time of observation.
30	31	32	33	34	35	36	37
Slight or moderate duststorm or sandstorm, has decreased during past hour.	Slight or moderate duststorm or sandstorm, no appreciable change during past hour.	Slight or moderate duststorm or sandstorm, has increased during past hour.	Severe duststorm or sandstorm, has decreased during past hr.	Severe duststorm or sandstorm, no appreciable change during past hour.	Severe duststorm or sandstorm, has increased during past hour.	Slight or moderate drifting snow, generally low.	Heavy drifting snow, generally low.
40	41	42	43	44	45	46	47
Fog at distance at time of ob., but NOT at station during past hour.	Fog in patches.	Fog, sky discernible, has become thinner during past hour.	Fog, sky NOT discernible, has become thinner during past hour.	Fog, sky discernible, no appreciable change during past hour.	Fog, sky NOT discernible, no appreciable change during past hour.	Fog, sky discernible, has begun or become thicker during past hr.	Fog, sky NOT discernible, has begun or become thicker during past hour.
50	51	52	53	54	55	56	57
Intermittent drizzle (NOT freezing) slight at time of observation.	Continuous drizzle (NOT freezing) slight at time of observation.	Intermittent drizzle (NOT freezing) moderate at time of ob.	Continuous drizzle (NOT freezing), moderate at time of ob.	Intermittent drizzle (NOT freezing), thick at time of observation.	Continuous drizzle (NOT freezing), thick at time of observation.	Slight freezing drizzle.	Moderate or thick freezing drizzle.
60	61	62	63	64	65	66	67
Intermittent rain (NOT freezing), slight at time of observation.	Continuous rain (NOT freezing), slight at time of observation.	Intermittent rain (NOT freezing), moderate at time of ob.	Continuous rain (NOT freezing), moderate at time of observation.	Intermittent rain (NOT freezing), heavy at time of observation.	Continuous rain (NOT freezing), heavy at time of observation.	Slight freezing rain.	Moderate or heavy freezing rain.
70	71	72	73	74	75	76	77
Intermittent fall of snowflakes, slight at time of observation.	Continuous fall of snowflakes, slight at time of observation.	Intermittent fall of snowflakes, moderate at time of observation.	Continuous fall of snowflakes, moderate at time of observation.	Intermittent fall of snowflakes, heavy at time of observation.	Continuous fall of snowflakes, heavy at time of observation.	Ice needles (with or without fog).	Granular snow (with or without fog).
80	81	82	83	84	85	86	87
Slight rain shower(s).	Moderate or heavy rain shower(s).	Violent rain shower(s).	Slight shower(s) of rain and snow mixed.	Moderate or heavy shower(s) of rain and snow mixed.	Slight snow shower(s).	Moderate or heavy snow shower(s).	Slight shower(s) of soft or small hail with or without rain or rain and snow mixed.
90	91	92	93	94	95	96	97
Moderate or heavy shower(s) of hail†, with or without rain or rain and snow mixed, not associated with thunder.	Slight rain at time of ob.; thunderstorm during past hour, but NOT at time of observation.	Moderate or heavy rain at time of ob.; thunderstorm during past hour, but NOT at time of observation.	Slight snow or rain and snow mixed or hail† at time of observa., thunderstorm during past hour, but NOT at time of observation.	Mod. or heavy snow, or rain and snow mixed or hail† at time of ob.; thunderstorm during past hour, but not at time of observation.	Slight or mod. thunderstorm without hail†, but with rain and/or snow at time of ob.	Slight or mod. thunderstorm, with hail† at time of observation.	Heavy thunderstorm, without hail†, with rain and/or snow at time of observation.

6-9 Reading the daily weather map.

Do You Know...... 111

ww		C_L Clouds of type C_L	C_M Clouds of type C_M	C_H Cloud of type C_H	C Type of cloud	W Past weather	N Total amount all clouds	a Barometer characteristic
08 Well developed dust devil(s) within past hr.	09 Duststorm or sandstorm within sight of or at station during past hour.	0 No Sc, St, Cu, or Cb clouds.	0 No Ac, As or Ns clouds.	0 No Ci, Cc, or Cs clouds.	0 St or Fs.	0 Clear or few clouds.	0 No clouds.	0 Rising then falling. Now higher than, or the same as, 3 hours ago.
18 Squall(s) within sight during past hour.	19 Funnel cloud(s) within sight during past hour.	1 Cu with little vertical development and seemingly flattened.	1 Thin As (entire cloud layer semitransparent).	1 Filaments of Ci, scattered and not increasing.	1 Ci	1 Partly cloudy (scattered) or variable sky.	1 Less than one-tenth or one-tenth.	1 Rising, then steady; or rising, then rising more slowly. Now higher than, or the same as, 3 hours ago.
28 Fog during past hour, but NOT at time of ob.	29 Thunderstorm (with or without precipitation) during past hour, but NOT at time of ob.	2 Cu of considerable development, generally towering, with or without other Cu or Sc; bases all at same level.	2 Thick As, or Ns.P.	2 Dense Ci in patches or twisted sheaves, usually not increasing.	2 Cs	2 Cloudy (broken) or overcast.	2 Two- or three-tenths.	2 Rising unsteadily, or unsteady. Now higher than, or the same as, 3 hours ago.
38 Slight or moderate drifting snow, generally high.	39 Heavy drifting snow, generally high.	3 Cb with tops lacking clear-cut outlines, but distinctly not cirriform or anvil-shaped; with or without Cu, Sc, or St.	3 Thin Ac; cloud elements not changing much and at a single level.	3 Ci, not anvil-shaped, derived from or associated with Cb.	3 Cc	3 Sandstorm, or duststorm, or drifting or blowing snow.	3 Four-tenths.	3 Rising steadily, or steady. Now higher than, or the same as, 3 hours ago.§
48 Fog, depositing rime, sky discernible.	49 Fog, depositing rime, sky NOT discernible.	4 Sc formed by spreading out of Cu; Cu often present also.	4 Thin Ac in patches; cloud elements continually changing and/or occurring at more than one level.	4 Ci, often hook-shaped, spreading over the sky and usually thickening as a whole.	4 Ac	4 Fog, or smoke, or thick dust haze.	4 Five-tenths.	4 Falling or steady, then rising; or rising, then rising more quickly. Now higher than, or the same as, 3 hours ago.
58 Drizzle and rain slight.	59 Drizzle and rain, moderate or heavy.	5 Sc not formed by spreading out of Cu.	5 Thin Ac in bands or in a layer gradually spreading over sky and usually thickening as a whole.	5 Ci and Cs, often in converging bands, or Ci alone, the continuous layer not reaching 45° altitude.	5 As	5 Drizzle.	5 Six-tenths.	5 Falling, then rising. Now lower than 3 hours ago.
68 Rain or drizzle and snow, slight.	69 Rain or drizzle and snow, mod'te or heavy.	6 St or Fs or both, but not Fs of bad weather.	6 Ac formed by the spreading out of Cu.	6 Ci and Cs, often in converging bands, or Cs alone, the continuous layer exceeding 45° altitude.	6 Sc	6 Rain.	6 Seven- or eight-tenths.	6 Falling, then steady; or falling, then falling more slowly. Now lower than 3 hours ago.
78 Isolated starlike snow crystals (with or without fog).	79 Ice pellets (sleet, U. S. definition).	7 Fs and/or Fc of bad weather (scud) usually under As and Ns.	7 Double-layered Ac or a thick layer of Ac, not increasing, or As and Ac both present at same or different levels.	7 Cs covering the entire sky.	7 Ns	7 Snow, or rain and snow mixed, or ice pellets (sleet).	7 Nine-tenths or overcast with openings.	7 Falling unsteadily, or unsteady. Now lower than 3 hours ago.
88 Moderate or heavy shower(s) of soft or small hail with or without rain or rain and snow mixed.	89 Slight shower(s) of hail†, with or without rain or rain and snow mixed, not associated with thunder.	8 Cu and Sc (not formed by spreading out of Cu) with bases at different levels.	8 Ac in the form of Cu-shaped tufts or Ac with turrets.	8 Cs not increasing and not covering entire sky; Ci and Cc may be present.	8 Cu or Fc	8 Shower(s).	8 Completely overcast.	8 Falling steadily. Now lower than 3 hours ago.§
98 Thunderstorm combined with duststorm or sandstorm at time of ob.	99 Heavy thunderstorm with hail† at time of ob.	9 Cb having a clearly fibrous (cirriform) top, often anvil-shaped, with or without Cu, Sc, St, or scud.	9 Ac of a chaotic sky, Ac of a chaotic sky, usually at different levels, patches of dense Ci are usually present also.	9 Cc alone or Cc with some Ci or Cs, but the Cc being the main cirriform cloud present.	9 Cb	9 Thunderstorm, with or without precipitation.	9 Sky obscured.	9 Steady or rising, then falling; or falling, then falling more quickly. Now lower than 3 hours ago.

6-10 Reading the daily weather map (continued)

7

Interpreting Weather Data

ONCE WEATHER INFORMATION IS GATHERED IT MUST BE ANALYZED IN ORDER TO BE USEFUL. Weather data may tell you that a high-pressure system is moving in your direction at about 75 miles a day. It is now about 200 miles from you and is nearly 300 miles wide. This data can be interpreted to say that in about two and a half days you can expect fair weather and warmer temperatures lasting about four days. To use this weather information, you might plan a beach outing during the time the high is in your area.

Interpreting weather data is as much a science as collecting weather information, though not as exact. Still, the rules of understanding and translating weather data into useful forecasts can easily be applied by the amateur meteorologist.

Weather is caused by the movement of the atmosphere as the Earth rotates around the Sun. Actually, it's caused by the movement of systems or areas of high and low-pressure. Chapter 2 gave you a background of how air masses, pressure systems, and fronts are formed and move.

By keeping the weather maps that appear in the daily newspaper and by watching the local television meteorologists you can keep accurate track of the surface pressure systems that move into, through, and out of your local area (FIG. 7-1). To better understand and forecast local weather, let's consider the air masses that affect weather in the continental United States.

MARITIME POLAR AIR MASSES

Maritime polar air masses arrive in the United States from two different source regions. One of these is located in the North Pacific Ocean, the other in the

114 INTERPRETING WEATHER DATA

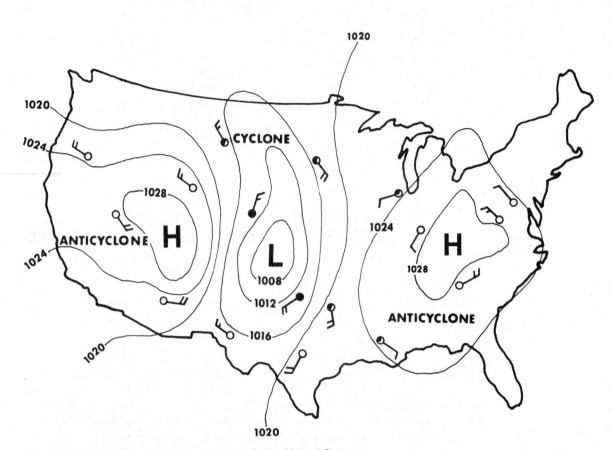

7-1 Surface windflow in pressure systems in the United States.

northwestern portion of the North Atlantic Ocean. The air masses from the Pacific Ocean affect the weather conditions of the Pacific coast of the United States and western Canada. The air masses that began over the North Atlantic Ocean appear during the winter over the northeastern coast of the U.S.

Many of the winter maritime polar air masses that arrive in the Pacific coast start in Siberia in the Soviet Union. They have a long overwater path. By the time they get here they are unstable in the lower layers (FIG. 7-2). As they invade the west coast, they are cooled from below by a cool ocean current and coastal area, and become more stable. Along the Pacific coastal regions, stratus and stratocumulus clouds are common in these air masses. Maritime polar air masses cause heavy cumuliform cloud groups and shower activity as they move eastward up the slopes of the mountains. East of the mountains, the air descends and is warmed by increasing pressure. This warming results in decreased relative humidity and the skies are generally clear.

In the northeastern United States, maritime polar air moves into the New England States from the northeast. These air masses are usually colder and more stable than those entering the west coast from a northwest direction. Low

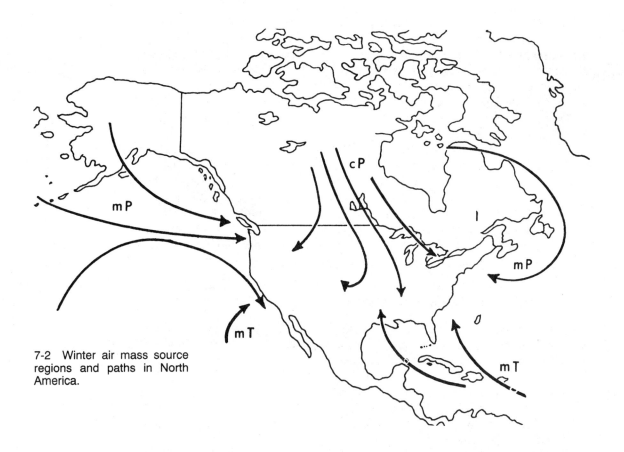

7-2 Winter air mass source regions and paths in North America.

stratiform clouds with light continuous precipitation and generally strong winds occur as these air masses move inland.

Since water temperatures are cooler than nearby land temperatures in the summer, maritime polar air masses entering the Pacific coast become unstable because of the surface heating. In the afternoon, cumuliform clouds cause widely scattered showers. At night fog and low stratiform clouds are common on the coastal regions, especially along the coast of California. When the air masses cross the mountains, they lose much of their moisture on the western mountain slopes in the form of heavy showers.

CONTINENTAL POLAR AIR MASSES

Continental polar air masses that invade the United States during winter start over Canada and Alaska. They are stable in these source regions. As the air masses move southward into the U.S., they are heated by the underlying surface. During daylight hours, the air is generally unstable near the surface and the sky is usually clear. At night the air tends to become more stable. When these cold, dry air masses move over the warmer waters of the Great

Lakes, they acquire heat and moisture and become unstable in the lower levels. Cumuliform clouds form and produce snow flurries over the Great Lakes and on the leeward side of the lakes. As the air masses move southeastward, the cumuliform clouds intensify along the Appalachian Mountains. Continental polar air masses between the Great Lakes and the peaks of the Appalachians contain some of the most unfavorable flying conditions in the United States during the winter months. Clear skies or scattered cloud are normal east of the mountains.

Cold dry air masses have different features in the summer than in the winter. Since the thawed-out source regions are warmer and contain more moisture, the air is less stable in the surface layers. The air is cool and contains slightly more moisture when it reaches the United States. Scattered cumuliform clouds form during the day in this unstable air, but scatter at night when the air becomes more stable. When these air masses move over the colder water of the Great Lakes in the summer, they are cooled from below and become stable, resulting in good weather.

MARITIME TROPICAL AIR MASSES

Maritime tropical air masses begin over the Atlantic Ocean, the Gulf of Mexico, the Caribbean Sea. They move into the U.S. from the Gulf of Mexico or Atlantic Ocean and are common along the Southeastern and Gulf Coast States (FIG. 7-3). Warm, moist, stable air that starts over the Pacific Ocean is rarely seen in Southwestern United States because the prevailing winds in the southwest blow offshore. Because the land is colder in winter than the water; warm, moist air masses are cooled from below and become stable as they move inland over the South Atlantic and Gulf States.

Fog and stratiform clouds form at night over the coastal regions. The fog and clouds often break up or become stratocumulus during the afternoon. The amount of cloudiness and fog that spreads inland depends on the air temperature. When surface temperatures are cold, fog and stratiform clouds extend inland for considerable distances throughout the Eastern States. When land temperatures are extremely cold, surface temperature inversions develop. Daytime heating usually doesn't eliminate the inversions, so the fog and stratiform clouds may remain for several days. In winter, when the air moves over the Appalachian Mountains, the cooling produced by orographic lifting makes heavy cumuliform clouds form on the windward side.

Warm moist air covers the eastern half of the United States during most of the summer. Since the land is normally warmer than the water, this air mass is heated from below by the surface and becomes unstable as it moves inland. Along the coastal regions, stratiform clouds are common during the early morning hours. These stratiform clouds usually change later in the morning to scattered cumuliform clouds. By late afternoon, wide areas of scattered thunderstorms normally develop. In maritime tropical air masses, cumuliform clouds and thunderstorms are usually more numerous and intense on the windward side of mountain ranges.

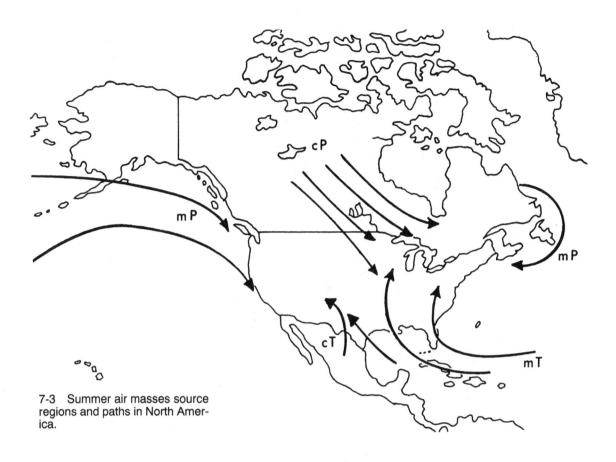

7-3 Summer air masses source regions and paths in North America.

CONTINENTAL TROPICAL AIR MASSES

Continental tropical air masses are seen in the Mexico/Texas/Arizona/New Mexico area, their source region, only in the summer. These air masses cause high temperatures, low humidity and sometimes even scattered cumuliform clouds. The bases of these clouds are very high, causing turbulent air below them.

PLOTTING THE WEATHER

Once you know the typical movement of air masses in your region of the country you can better predict local weather (FIG. 7-4). These weather systems move at average speeds of 15 miles an hour in the summer to 25 miles an hour in winter months and are carried along by air streams called the *prevailing westerlies*. Information gathered on barometric pressure will tell you whether you are now within a high-pressure, above 1,016 millibars or 30 inches of mercury, or a low-pressure, below 1,016 millibars, system. Barometric pressure that is rising or falling can tell you whether the center of the pressure system is coming closer or moving away from your location.

118 INTERPRETING WEATHER DATA

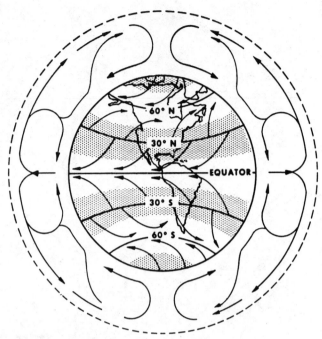

7-4 Three cell theory of circulation.

Pressure above 30 inches of mercury and rising means the high-pressure system is coming toward you. Falling pressure means it is moving away. Pressure readings below 30 inches of mercury and falling mean a low-pressure system is coming in. Rising pressure would mean it has passed.

You can also check the wind to decide what type of pressure system it is. Wind flows more or less with the isobars, the lines connecting areas of same barometric pressure, clockwise around a high and counterclockwise around a low pressure system. A simple way of finding high and low systems in your area is to turn your back on the wind, then point to the left toward the low pressure system and right for the high pressure system (FIG. 7-5). Another rule of thumb is that highs often bring good weather and lows bring bad weather.

You can then plot the weather by using newspaper or television weather maps to compare where a system was yesterday and where it is today. Note how many miles the system has moved in the past 24 hours, then forecast where the system will be tomorrow. Keep two things in mind as you plot. First, the earth rotates which spins systems to the left or right of a straight path depending on the type of system. Second, most maps are flat maps rather than round and distort the actual distances and paths that systems seem to take.

FORECASTING YOUR OWN WEATHER

Since these high and low-pressure systems bring in the weather to an area, the ability to read incoming systems in advance will help you predict tomorrow's weather. And, as we have just seen, two excellent clues of future systems are

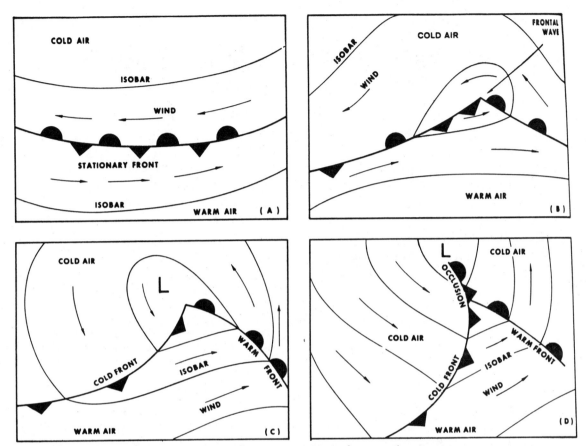

7-5 Stages in the development of the occluded wave on a surface weather map.

the barometric pressure, highs and lows, and the wind direction, location and path of the system. If you keep a chart of wind direction and barometric pressure relationships, it can be used to forecast weather on a short-term basis. Table 7-1 is such a chart, useful for many parts of the United States. With this chart, a weather-vane, and a barometer you can predict weather for up to 48 hours in advance with a high degree of accuracy.

For example, your barometer reads 30.10 to 30.20 inches of mercury and falling slowly while the wind comes from the south or southwest. You can probably expect rain within 24 hours. Why? Because a low-pressure system is moving in, pushing out the high.

You can also confirm your barometric/wind forecasts by reading the clouds. Clouds mark the circulation of the lower atmosphere, trailing out along the bands of wind at various levels, or growing with vertical movement of the air. Because clouds are made up of condensed liquid or frozen water, their visible features also reflect temperature and moisture content of the atmosphere in the area.

The cirrus or ice crystal cloud, for example, is generally followed by low and middle clouds. A cross-sectional view along this parade of clouds would show them forming along a slope, ranging from cirrus at the high end to altostratus or

Table 7-1 Simple Wind/Pressure Weather Forecasting Chart

Wind direction	Barometer reduced to sea level	Character of weather indicated
SW. to NW	30.10 to 30.20 and steady	Fair, with slight temperature changes for 1 to 2
SW. to NW	30.10 to 30.20 and rising rapidly	Fair, followed within 2 days by rain
SW. to NW	30.20 and above and stationary	Continued fair, with no decided temperature change.
SW. to NW	30.20 and above and falling slowly	Slowly rising temperature and fair for 2 days
S. to SE	30.10 to 30.20 and falling slowly	Rain within 24 hours.
S. to SE	30.10 to 30.20 and falling rapidly	Wind increasing in force, with rain within 12 to 24 hours.
SE. to NE	30.10 to 30.20 and falling slowly	Rain in 12 to 18 hours.
SE. to NE	30.10 to 30.20 and falling rapidly	Increasing wind, and rain within 12 hours.
E. to NE	30.10 and above and falling slowly	In summer, with light winds, rain may not fall for several days. In winter, rain within 24 hours.
E. to NE	30.10 and above and falling rapidly	In summer, rain probable within 12 to 24 hours. In winter, rain or snow, with increasing winds, will often set in when the barometer begins to fall and the wind sets in from the NE.
SE. to NE	30.00 or below and falling slowly	Rain will continue 1 to 2 days.
SE. to NE	30.00 or below and falling rapidly	Rain, with high wind, followed, within 36 hours, by clearing, and in winter by colder.
S. to SW	30.10 or below and rising slowly	Clearing within a few hours, and fair for several days.
S. to E	29.80 or below and falling rapidly	Severe storm imminent, followed within 24 hours, by clearing, and in winter by colder.
E. to N	29.80 or below and falling rapidly	Severe northeast gale and heavy precipitation; in winter, heavy snow, followed by a cold wave.
Going to W	29.80 or below and rising rapidly	Clearing and colder.

stratus at the other. Turbulence, unstable air in air masses, can also be read by the type of cloud. Approaching cirrocumulus clouds predict unstable air.

Remember that improving weather conditions can be forecast by a steady decrease in the number of clouds, higher cloud bases, increasing breaks in the overcast, and fog that scatters before noon. Poor weather can often be forecast if fast-moving clouds thicken and lower, a line of middle clouds darken on the western horizon, isolated rollclouds fuse into sheet clouds, and lower or darker clouds develop dark bases.

HOW THE NWS FORECASTS WEATHER

The National Weather Service uses the latest in meteorological and computer technology to forecast tomorrow's weather from data gathered by observers and satellites. Let's consider their methods and how you can use them to forecast your local weather.

The most common method of forecasting weather, for both the amateur and professional meteorologist, is known as *syntopic forecasting*. It uses a summary of the total weather picture to forecast into the future. The movement of weather systems is illustrated on a sequence of weather maps, also called syntopic charts. Observations noted on these maps are made at thousands of weather stations around the world many times each day (FIGS. 7-6 and 7-7). Information on the upper atmosphere is also charted daily to help forecast weather conditions in your area.

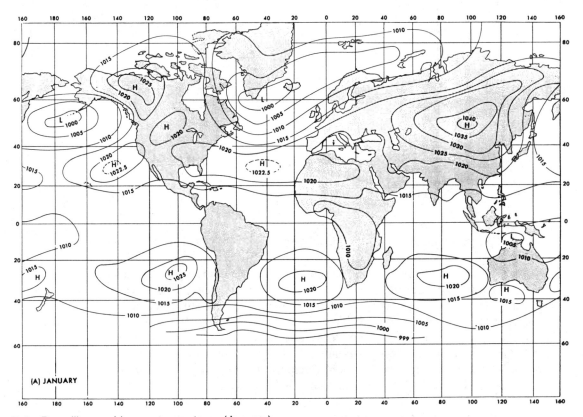

7-6 Prevailing world pressure systems (January).

Another method of reading weather, called *statistical forecasting*, uses mathematical equations based on past weather patterns. It says that, since yesterday's temperature was 78, the temperature a year ago was 82, and the normal

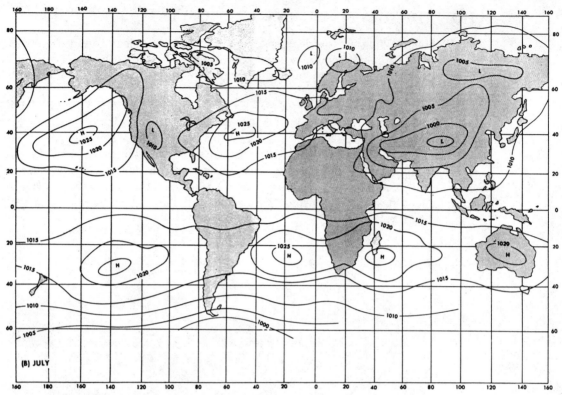

7-7 Prevailing world pressure systems (July).

over the last 12 years for this date is 80, that the high today will be about 80 degrees.

Numerical forecasting uses mathematical models based on the physical laws of the Earth's atmosphere using the science of fluid dynamics. In theory, complete data on the Earth's atmosphere, water bodies, and land surfaces, plus a complete understanding of the physical laws of heat and moisture transfer could yield almost perfect numerical weather forecasts. Unfortunately, such information isn't fully available. However, the advent of high-speed computers in the last few decades is changing this situation. Computers at the National Meteorological Center in Suitland, Maryland, compute the weather future in 20 minute intervals.

Beyond about 90 days, weather can be predicted just as well through *climatological forecasting* using the averages of past weather records. This method is also very accurate for general weather conditions. After syntopic weather forecasting, climatological forecasting is most often used by the amateur meteorologist for predicting weather.

CLIMATOLOGY

Climatology is the study of climate, the normal weather for an area. Climatology's job is to describe and then predict climate by gathering and analyzing longterm weather data. The basic elements of weather (temperature, pressure, humidity, precipitation, winds, clouds, sunshine, and visibility) and storms (hurricanes, tornadoes, and thunderstorms) are seen and recorded. These records are used to determine mean or average extremes, totals, and probabilities of weather.

CONSIDERING MICRO-CLIMATES

Temperatures may be very different from point to point over short distances, especially in areas with mountains. These variations are caused by differences in altitude, slope of land, type of soil, plants, bodies of water, air drainage, and urban heat effects. These conditions bring about *micro-climates*.

An increase in altitude of 1000 feet generally causes a decrease of 3.3°F in the average annual temperature. An extreme case exists in the Grand Canyon area of Arizona between Inner Canyon, elevation 2500 feet, and Bright Angel, elevation 8400 feet — 5900 feet higher but just five miles away. The normal daily highest and lowest temperatures in July are 106°F and 77°F at Inner Canyon, compared with 78°F and 47°F at Bright Angel; in January it's 55°F and 37°F against 37°F and 15°F.

Because of the daily motion of the Sun, an eastern slope will be warmer in the mornings and colder in the afternoons than a western slope. A southern slope will remain warmer all day, while a northern slope will remain cool.

At night, especially on clear and calm nights, air near the ground cools because of radiative cooling. The cold air is heavier so if the land is sloping, the cold air will drain off and flow to lower elevations. At the original elevation the cold air will be replaced by warmer air. Farmers use this information and plant hardy vegetation in the bottom of the valleys and more tender vegetables and fruits in the warmer belts just above.

Bare, dry soil will heat and cool more rapidly than heavy, claylike, wet soil, or a soil covered with vegetation. Light colored surfaces will be cooler because they reflect a large portion of the incoming Sun's rays. Dark surfaces will be warmer because they absorb much of the solar heat.

Bodies of water, such as the Atlantic and Pacific Oceans, the Gulf of Mexico and the Great Lakes have a moderating effect on nearby land areas. This effect varies depending on wind, land shape, and size of the water area.

Cities and industrial areas usually average a few degrees warmer with less extreme temperatures than the nearby countryside. This is due to the trapping of solar heat in the city by buildings, sidewalks, and streets. Heat produced in domestic and industrial furnaces, stoves, and incinerators, is held near the surface by the blanketing effects of smoke, gases, dust, and other pollutants in the air over the city.

The prevailing winds over an area are a strong influence on the type of climate the area will have. Onshore winds over coastal areas usually give such ar-

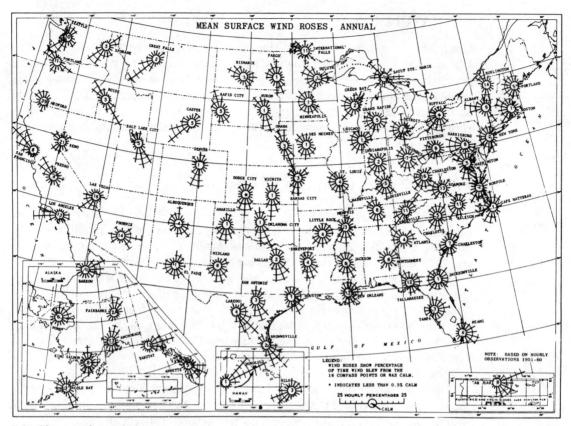

7-8 Mean surface wind roses, annual.

eas a milder and moister climate than if the winds were from the interior of the continent. The prevailing winds on the Pacific coast are from the northwest (FIG. 7-8). Winds over the eastern two-thirds of the United States are from the northwest or north during January and February, and generally from south to southwest from May through August. In the Texas/Oklahoma area, however, southerly winds prevail from March through December. Easterly winds of the northeast trades prevail over the Florida peninsula except during December and January when the northern winds bring cooler weather.

Relative humidity is a percentage of the amount of water vapor in the air and the amount of water vapor the air could hold at a given temperature. The higher the temperature, the more water vapor the air can hold. An increase of 20°F usually doubles the capacity of air to hold water vapor. So when air temperature increases 20°F, an existing 100 percent relative humidity will decrease to 50 percent. In general, relative humidities are lower during the afternoon when the air temperature reaches its daily maximum, and higher in the early morning when the minimum temperature of the day occurs.

IMPROVING FORECAST ACCURACY

We can all understand the feelings of the angry woman who called up the local weather office early one morning and complained, "Ya wanna get someone out here to shovel off this six inches of 'partly cloudy'?" Weather forecasts are not always accurate. In fact, sometimes they can be 180 degrees off. Most people have been to picnics where it rains, heard reports of thunderstorms that never appeared, and suffered wind storms that caught them unprepared.

Unfortunately, meteorology is not an exact science. There are too many elements that play a part in the weather that reaches our location. An unexpected change in a high altitude wind may veer a low-pressure system away from us by only 50 miles and give us crystal clear weather as we pull out umbrellas. Or a distant front may alter the local front and modify the path just a few miles, enough to miss you. Or a storm may weaken before reaching you and pass harmlessly overhead.

The major cause of inaccurate weather forecasts is the need to generalize. That is, a regional forecast may state that weather will be partly cloudy, but a high range of mountains may condense the clouds as they try to rise above it. The heavier clouds now climb the slope, getting colder and colder until they drop six inches of "partly cloudy" onto a home near the summit. On the other side of the summit, a neighbor complains about the dry weather. These are the microclimates discussed earlier.

So how can you improve the accuracy of weather forecasts for your microclimate? First, by recognizing the conditions around you that are not "normal" or typical in your area: mountains in a desert, a large lake near a city, a low valley surrounded by mountains, or a large forest nearby.

Then you must consider what affect these elements will have on the weather. Will they increase the temperature? Decrease it? Moderate temperature extremes? Add relative humidity? Reduce winds? Cause tornadoes or waterspouts? Weaken thunderstorms? You can do this by reviewing the early chapters of this book or talking with other amateur meteorologists. Or you can learn by observation and comparison. When the weatherman predicts scattered showers by noon and you don't receive them until nearly two hours later, mark it down in your weather log. Long-term temperature comparisons may show that your local temperature is 2°F warmer and 3°F cooler than the "official" air temperatures offered on the weather radio.

Your weather log may also offer information on other differences between your weather and the official weather for your area that is often taken at an airport. Your micro-climate may offer more or less precipitation, faster or slower winds, higher or lower relative humidity, or other elements. A comparison can be useful in revising the official weather forecast to be more accurate for your own corner of the world.

By collecting accurate data and adding the resources of the NWS satellites, computers, and professional meteorologists, you can make highly accurate forecasts of local weather hours, days, and even months in advance. In fact, your forecasts can be even more accurate for your location than those issued by the Weather Service. You can be the local "weatherman."

DO YOU KNOW...

- The difference between a *maritime* and *continental* polar air mass?
- The difference between a *tropical* and a *polar* air mass?
- What a falling barometric pressure means?
- What type of forecasting looks beyond 90 days?
- What a *micro-climate* is?
- How to forecast weather even better than the National Weather Service?

8

Using the Weather

It was early on the morning of June 18, 1815. All night long, heavy rain, lightning, thunder, and high winds had softened the ground around the two armies. Rather than attack at dawn, one general decided to wait for the Sun to come out and dry the ground so the artillery could be moved forward. The Sun never came out that day, but the wait gave the enemy's reinforcements enough time to attack the general from the rear, causing the hasty retreat of Napoleon and his forces at Waterloo. The war was lost.

On the same continent 129 years later, another general worked his climatologists around the clock to predict the ideal conditions for his air, sea, and land forces to overrun the beach at Normandy. General Dwight D. Eisenhower used the weather to his advantage and turned the tide of World War II.

Today, weather forecasts by amateur and professional meteorologists are used in every part of daily life from raising crops to flying airplanes to planning picnics to waging war. Weather information can be used to plan baseball games and department store sales, help select horserace winners, buy the best fruits and vegetables, and even predict crime waves. Weather affects every part of our lives whether we understand it or not. Our knowledge of weather can help us appreciate and use the treasures of this world.

AGROMETEOROLOGY

Agrometeorology is a 60-cent word that means using the weather to help in planting, raising, and harvesting food. It's the study of the partnership between weather and agriculture.

Of course, in growing crops, the most important weather element is precipitation. Water is needed in the right amount to ensure adequate moisture for growth, not too much and not too little. Whether you're raising 640 acres of soybeans or a 100-square-foot plot of vegetables, the amount of moisture your "farm" receives is important to its location, care, and selection of crops.

Growers depend on both long-term and short-term weather forecasts. They want to know if the growing season will be long enough for sugar beets, but they also need to know if their 80 acres of feeder corn are threatened by hail. The key to these problems is in weather records. What do the climatologists say are the growing season or the chance of hail?

The freeze-free growing period is between the dates of the last 32°F temperature in the spring and the first 32°F temperature in autumn. It is usually referred to as the "growing season". Temperatures, however, may drop to freezing or below at the ground surface because of radiation on clear, calm nights, even when the official low temperature measured in the weather shelter is as high as 10°F above freezing. Furthermore, very sensitive plants will be damaged or killed by temperatures as high as 37 to 42°F. The growth rate of very hardy vegetation, such as winter cereals and many grasses, usually is only restricted by freezing temperatures. Such vegetation remains dormant during longer cold weather spells and starts to grow again when moderate warmth returns.

The effect of low temperatures on vegetation depends on the variety and condition of plants before the freeze, the length of time that low temperatures continue, and protection offered by a blanket of snow or other covering. Winter wheat, for instance, is economical to grow in the Great Plains as far north as north central South Dakota. It will freeze during many winters if planted farther north. However, winter wheat is grown successfully farther north in the sheltered snow-covered valleys of western Montana. A growing season of at least 180 to 190 days is needed for the growth of cotton, 120 to 130 days for grain corn, and 90 to 100 days for spring wheat. The long hours of summer daylight in northern areas can make up for the short growing season. For average world temperatures, see FIGS. 8-1 and 8-2.

Many other temperature factors determine the location and type of crops the farmer or gardener sows. One such factor is the frequency of a "hard freeze," usually early in autumn or late in spring. Apples, peaches, cherries, and grapes are grown along the Great Lakes, mainly on the leeward (eastern and southeastern) side because of the moderating effects of the lake waters on air masses passing over them. This moderating effect slows down the too-early growth of fruits in spring and reduces the chances of late frosts in spring and early frosts in autumn. On a larger scale, the moderating effect of the Pacific Ocean helps make the Pacific States one of the greatest fruit and vegetable regions of the world.

THE FARM WEATHERFORECASTER

In this age of special jobs, you might think that the big business of agriculture has its own specialized meteorologists. It does. In fact, there are many men and women who make a full-time living telling farmers what they can expect over the horizon tomorrow, next month, and even next year. They can also help the com-

THE FARM WEATHERFORECASTER 129

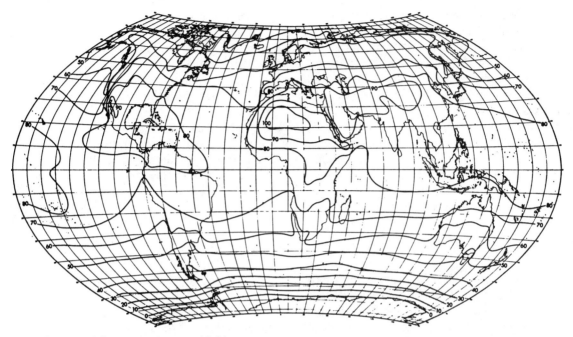

8-1 Average July temperature worldwide.

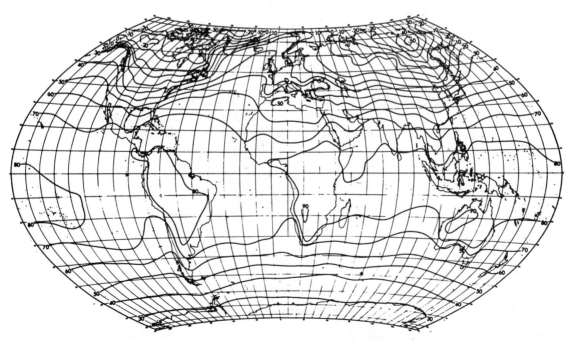

8-2 Average January temperature worldwide.

mercial farmer decide when and how to sell his crop, because weather also plays a major part in food marketing.

Some farmers work directly with these agrometeorologists to plan crops and marketing. Others subscribe to printed and recorded agriculture weather services. Still others serve as their own weather forecasters by setting up a weather station and keeping a weather log or combining weather entries into their farm operations log.

The home gardener can improve crop yield and weather forecasting. With the help of a garden shop, the home gardener can select "early" and "late" crops that fit the local and micro-climate growing season. The gardener can also estimate watering needs to decide which food crops will cost the least to grow. In many cases, the gardener or farmer can get help on local weather and crops through the County Extension Service.

WEATHER AND THE PILOT

The second largest user of weather information is the private pilot. Commercial aircraft often fly will above the weather and only need to consider winds aloft and the weather where they take off and land. But the pilot of the small airplane is surrounded by the weather. The wind governs his true airspeed. The visibility and cloud cover rules his flying altitude and whether or not he can fly. In spite of the excellent weather forecasting services available to general aviation pilots, the weather remains a major cause of accidents. The pilot worries about lack of visibility, turbulence, icing, hail, as well as strong surface winds.

The visibility problem may be caused by one of several kinds of fog, clouds, storms, snow, haze, smog, and dust. The most dangerous turbulence is found in or near thunderstorms, low-altitude wind shears, mountains, and in the wakes of the large aircraft.

There are two types of aircraft icing: structural and powerplant. Structural icing may be expected anytime the free temperature is 0°C (32°F) or colder and the humidity high. Carburetor ice may form in conditions of high humidity with temperatures as low as 10°C and as high as 25°C. It is most serious when the temperature and dewpoint approach 20°C. Air intake ducts are most likely to ice when the temperature and dewpoint are near 10°C or less.

The greatest danger from hail outside of a thunderstorm is directly beneath a thunderstorm and in clear air beneath the anvil top of a mature thunderstorm that tops 35,000 feet or more.

Strong surface winds cause problems when pilots are taxiing, landing, and taking off. Usually, these are local winds that create turbulence downwind from surface rises.

Dangerous low-altitude *wind shear* near airports is caused by a temperature inversion. This occurs when a layer of fast-moving warm air is above colder, relatively calm air next to the surface.

Let's take a closer look at each weather element and how it aids or makes life more difficult for the private pilot. Strap on your leather cap, adjust your goggles, and make one more wrap with your neck scarf.

Fog and Flying

Fog forms as moist air cools to its dewpoint, or as moisture is added to the air near the ground (FIG. 8-3). Ideal conditions for fog are high relative humidity, plenty of condensation particles, light surface winds, and a cooling process that starts the condensation. Fog is more common over coastal areas where moisture and particles are plentiful. However, fog often occurs in some industrial areas inland even when humidity is less than 100 percent due to an abundance of particles in the air.

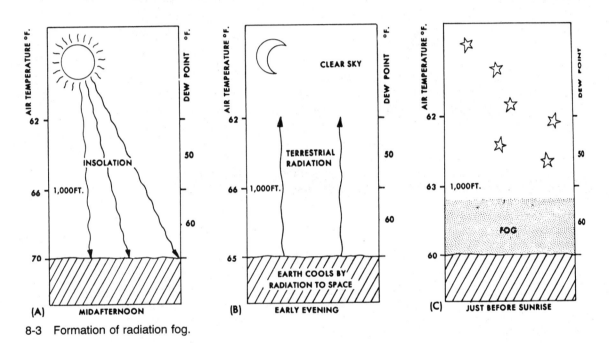

8-3 Formation of radiation fog.

Ground fog is actually called radiation fog. It forms on clear, calm nights when the ground cools the air to the dewpoint temperature. It usually is shallow, favors level ground, and generally burns off well before noon.

Advection fog is a coastal fog that forms when moisture air moves over colder ground or water. It can form side by side with ground fog. When the surface wind is greater than 15 knots, it will lift to form a layer of low stratus clouds.

Stream fog forms over water when the movement of cold air over much warmer water sets up intense evaporation. Because stream fog forms over a warm surface, the heating from below makes the air unstable. Turbulence and icing may be expected with this kind of fog. It is often seen in the polar regions, but is sometimes found over lades and rivers in the northern United States during autumn.

Precipitation induced fog is caused by the addition of moisture to the air through evaporation of rain or drizzle. It is most frequently found near warm fronts, though it can occur independently. This kind of fog usually forms very quickly and covers a large area.

Chasing Clouds

The atmosphere's movement, water content, and stability can be read in the clouds. In stable air stratus clouds form in layers with little or no vertical development. Thickness ranges from a few hundred feet to several thousand. Precipitation is drizzle, continuous light rain, or snow. There's little or no turbulence with stratus formations. However, the danger of structural icing is great if the free air temperature is 0°C or colder. Carburetor ice is likely if temperature is 25°C or lower. So the presence of middle and/or low stratus clouds usually means poor flying weather.

Cumulonimbus clouds are also called "thunderheads." They are rarely larger than ten miles in diameter but often run in packs, passing out damaging weather for a hundred miles and lasting up to eight hours. Because they often reach from ground level to as high as 60,000 feet, the pilot can't go under, over, or around them. The only thing to do is sit them out on the ground. Figure 8-4 shows the development of these powerful cloud systems.

Squall lines often develop 50 to 300 miles ahead of fast-moving cold fronts, bringing severe weather such as heavy hail, destructive winds, tornadoes, and thunderstorms.

Turbulence

There are four main causes of air turbulence that make things rough on the pilot. They are vertically moving air in convective currents, air moving around or over mountains and other obstructions, wind shear, and the passage of large aircraft.

If you've flown in small aircraft you are familiar with convective currents that cause bumpiness in lower altitudes during warm weather. When enough moisture is present this process will produce cumulus clouds.

Turbulence that results from air near the surface flowing over rough terrain, trees, buildings, and hills may set up tricky situations for light planes during landing and takeoff operations. When the wind is light there is not much turbulence. In wind speeds above 20 knots, the flow may be broken up into irregular eddies that can stay downwind to create a hazard in the landing area.

Especially dangerous are the winds that blow up the windward slope of mountains. The air above them is fairly smooth in a stable atmosphere. However, as winds are strong, you need at least 2000 feet between you and the ridge for a reasonably safe crossing.

Wind shear occurs when a stream or mass of air moves at a high speed and usually in a direction different than the air mass directly adjacent to it. This condition is fairly common over the northern U.S. and in Canada during winter where temperature inversions occur near the surface and the terrain below, usually a valley, holds cold, calm air beneath a moving layer of warm air.

Wake turbulence acts like a pair of tiny twin tornadoes generated by the airfoils of large aircraft. These twisting columns of air stream behind the big

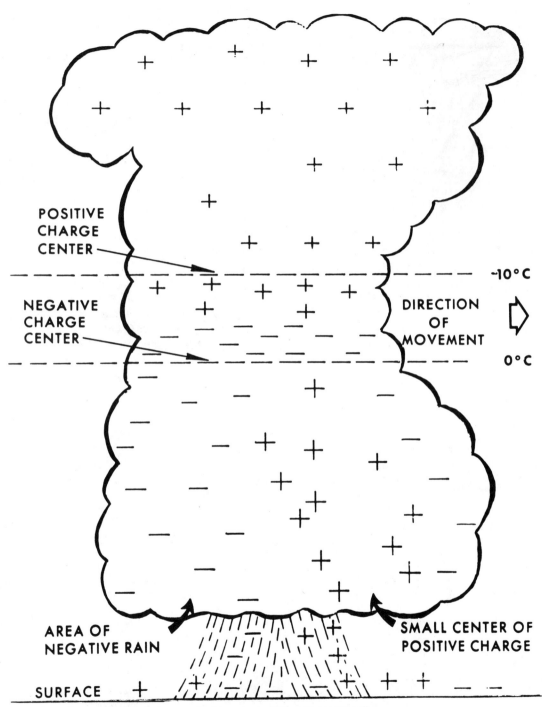
8-4 Location of electrical charges inside a typical thunderstorm cell.

planes from each wingtip with a downward force of about 1500 feet per minute and can damage light aircraft caught within one. Normally, wake turbulence lasts for only a minute or two after the passage of a large aircraft.

Aircraft Icing

Aircraft icing is a major weather hazard to aviation. Its formation can dangerously distort airfoil shapes, add drag and weight, and make the plane vibrate. Ice deposits of only one-half inch on the leading edge of wings on some aircraft reduce their lifting power as much as 50 percent. They also increase drag as well as the stall speed, at which a plane can no longer fly.

WEATHER AND BOATERS

Boating is a growing pastime in the United States and many new boaters discover the waters of our country each year. They, especially, must become aware of the weather and the part it plays in safe boating.

Figure 8-5 shows marine weather service charts available throughout the United States offering National Weather Service office telephone numbers and locations of warning display stations. These charts are available at local marinas and marine chart dealers or by ordering from Distribution Division (C44), National Ocean Survey, 6501 Lafayette Avenue, Riverdale, MD 20840.

Figure 8-6 shows display signals used by marine weather stations to show sea conditions. They are called *Small Craft Advisories* and cover a wide range of

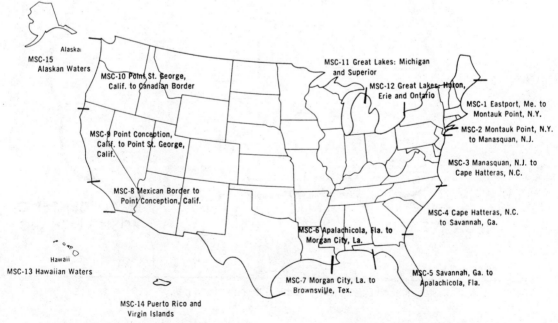

8-5 Location of marine weather service charts.

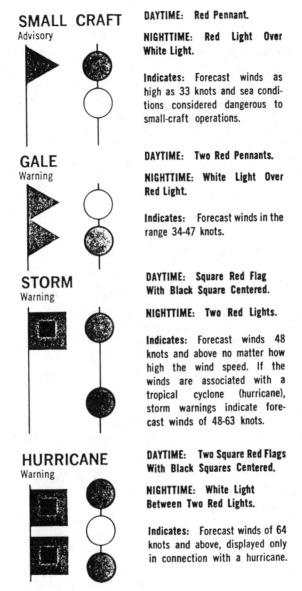

8-6 Marine weather advisory signals.

wind and sea conditions for boats of many sizes and designs. Hurricane warning signals are also offered.

The first step in safe boating is obtaining the latest available weather forecast for your boating area. Where they can be received, the NOAA Weather Radio broadcasts are the best way to keep informed of expected weather and sea conditions.

Keep a weather eye out for the approach of dark, threatening clouds that may forecast a squall or thunderstorm, any steady increase in wind or sea, or an increase in wind speed opposite in direction to a strong tidal current. A dangerous rip tide condition may form steep waves capable of capsizing a boat.

Of course, check radio weather broadcasts for the latest forecasts and warnings. Heavy static on your AM radio can warn you of a nearby thunderstorm. If a thunderstorm catches you while afloat, you should remember that in addition to gusty winds, lightning also poses a threat to safety. The best rules are to stay below deck if possible, keep away from metal objects that are not grounded to the boat's protection system, and don't touch more than one grounded object at the same time.

A bimonthly publication produced by the National Weather Service, The Mariner's Weather Log, offers articles on marine meteorology and climate. It is available through the Superintendent of Documents, U.S. Government Printing Office, Washington, DC 20402.

INDUSTRIAL METEOROLOGY

Business and industry also depend on the weather and can use it to their advantage in production and trade. Weather influences transportation by truck, train, barge, ship, and plane. Weather, in the form of heavy storms, can also reduce or completely stop production through the loss of the labor force for a few days or weeks. Rain, snow, and high humidity can damage raw and finished products left out in the open. Weather can also play an important part in how wastes are discharged from a factory. Inversion layers may not allow wastes to be put into the air. A lack of snow run-off can reduce river levels and force industries to find alternative and more expensive methods of handling wastes.

Even safety is relative to weather. In coal mines, gas flows out of the coal when the barometric pressure lowers. This increases the chances of explosions within the mine. And, as you learned in Chapter 4, the weather can change human personalities for the better or worse, changing how well people work.

WEATHER AND COMFORT

One of the most useful weather statistics is the *heating degree day*, the number of degrees the average temperature for the day is below 65°F. For example, a day with an average temperature of 50°F has 15 heating degree days. A day with an average temperature of 65°F or higher has no heating degree days. The amount of heat required at lower temperatures relates to the degree day value. A fuel bill usually would be about twice as high for a month with 1000 heating degree days as for a month with 500 heating degree days.

In contrast, *cooling degree days* are the number of degrees the average temperature for the day is above 65°F.

More important to human comfort is the effective temperature. The *effective temperature* is an index of the degree of warmth felt by the body on exposure to different combinations of temperature, humidity, and air movement. This measurement was developed in a series of studies in 1923 at the Research Laboratory of the American Society of Heating and Air Conditioning Engineers. Figure 8-7 shows how effective temperature relates to comfort.

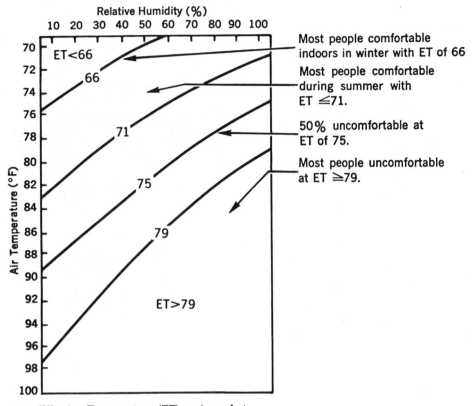

8-7 Effective Temperature (ET) and comfort.

The temperature-humidity index is a measurement developed in 1958 as a modification of the effective temperature method. The temperature-humidity index or *THI*, can be computed from one of the following equations:

$$THI = 0.4(t_d + t_w) + 15$$

$$THI = 0.55 t_d + 0.2 t_{dp} + 17.5$$

$$THI = t_d - [(0.55 - 0.554 RH)(t_d - 58)]$$

where t_d = dry bulb temperature
  t_w = wet bulb temperature
  t_{dp} = dewpoint temperature
  RH = relative humidity

All "t" values are in degrees Fahrenheit and the "t" values and the RH in each equation are simultaneous readings. The relative humidity is used in its decimal form (i.e., 0.65 instead of 65 percent).

A temperature-humidity index of 60 is generally used as a base for comparison. In summer, relatively few people will be uncomfortable because of heat and humidity while the index is 70 or below, about half the population will be uncom-

Table 8-1 Wind Chill Equivalent Temperature Chart

Wind speed (mi/h)	Dry bulb temperature (°F)																	
	45	40	35	30	25	20	15	10	5	0	−5	−10	−15	−20	−25	−30	−35	−40	−45
4	45	40	35	30	25	20	15	10	5	0	−5	−10	−15	−20	−25	−30	−35	−40	−45
5	43	37	32	27	22	16	11	6	0	−5	−10	−15	−21	−26	−31	−36	−42	−47	−52
10	34	26	22	16	10	3	−3	−9	−15	−22	−27	−34	−40	−46	−52	−58	−64	−71	−77
15	29	23	16	9	2	−5	−11	−18	−25	−31	−38	−45	−51	−58	−65	−72	−78	−85	−92
20	26	19	12	4	−3	−10	−17	−24	−31	−39	−46	−53	−60	−67	−74	−81	−88	−95	−103
25	23	16	8	1	−7	−15	−22	−29	−36	−44	−51	−59	−66	−74	−81	−88	−96	−103	−110
30	21	13	6	−2	−10	−18	−25	−33	−41	−49	−56	−64	−71	−79	−86	−93	−101	−109	−116
35	20	12	4	−4	−12	−20	−27	−35	−43	−52	−58	−67	−74	−82	−89	−97	−105	−113	−120
40	19	11	3	−5	−13	−21	−29	−37	−45	−53	−60	−69	−76	−84	−92	−100	−107	−115	−123
45	18	10	2	−6	−14	−22	−30	−38	−46	−54	−62	−70	−78	−85	−93	−102	−109	−117	−125

fortable when the index reaches 75, and almost everyone will be uncomfortable when it reaches 80.

WIND CHILL

The effect of a given temperature on the human body depends on the wind and humidity conditions. The effects of humidity were discussed in the section on Effective Temperature. The wind lowers body temperature by evaporating perspiration or by blowing heat away from the surface of the skin. Table 8-1 shows the temperature you feel on your skin under given conditions of temperature and wind. Note that the chilling effect increases as the wind speed increases. When winds exceed 40 miles per hour, there is little additional chilling effect.

USING DAILY WEATHER

The conditions of weather can be used in everyone's life, weather you are a student, farmer, gardener, flier, traveller, industrial plant operator, retail store manager, sports fan, teacher, or truck driver.

Weather can be applied through understanding the basics of how and why it forms in the atmosphere. We must also understand those elements that directly reach us, such as wind, pressure, temperature, humidity, precipitation, and clouds.

Weather is used by watching its patterns and using these rules to predict future weather conditions. The signs come from observation of the sky, nature, and ourselves. These observations are recorded in diaries and logs. Weather elements can be recorded using the simplest instruments you can make or find yourself, or with data from the National Weather Service and amateur meteorologists.

Weather data is then collected from its many sources and interpreted as "forecasts." Yet even the person with no weather instruments can often predict the weather in his or her area more accurately than the weatherman's Area Forecasts by observing and interpreting local weather conditions. All the amateur meteorologist needs is a weather radio, a simple thermometer or two, and an understanding of why and how weather works in our lives. This knowledge can then be applied to decide what clothing to wear, plan vacations, predict when business will increase or decrease, decide on appropriate outdoor activities throughout the year, prepare for bad weather, predict people's moods and energy levels, and take advantage of weather conditions.

Most important, learning how weather affects our daily lives can give each of us a greater appreciation for the Earth on which we live and its Creator.

DO YOU KNOW...

- Why weather forecasting is important to army generals?
- What an *agrometeorologist* does and why?
- How a farmer uses the weather to grow more crops?

140 USING THE WEATHER

- Why pilots of small aircraft must worry more about the weather than airline pilots?
- What *wake turbulence* is?
- How to figure the effective temperature for today?

Glossary

Active front A front that produces extensive cloudiness and precipitation.

Anemometer An instrument for measuring the force or speed of the wind.

Anvil cloud The popular name of a heavy cumulus or cumulonimbus cloud having an anvillike formation of cirrus clouds in its upper portions. If a thunderstorm is seen from the side, the anvil form of the cloud mass is usually noticeable.

Arctic front The zone of discontinuity between the extreme cold air of the Arctic regions and the cool polar air of the northern Temperate Zone.

Blizzard A violent, intensely cold wind laden with snow.

Buildup A cloud with considerable vertical development.

Ceiling The height above the Earth's surface of the lowest layer of clouds that is reported as broken, overcast or obscured and not as thin or partial.

Celsius scale A centigrade temperature scale originally based on 0°F for the boiling point of water, (i.e.), an inverted centigrade scale. It is now used interchangeably with the centigrade scale, with 0°C as the freezing point and 100°C as the boiling point.

Cloud band A mass of clouds, usually of considerable vertical extent, stretching across the sky on the horizon, but not extending overhead.

Cloudburst A sudden and extremely heavy downpour of rain; frequent in mountainous regions where moist air encounters orographic lifting.

Cold wave A rapid and marked fall of temperature during the cold season of the year. The National Weather Service applies this term to a fall of temperature in 24 hours equaling or exceeding a specified number of degrees and reaching a specified minimum temperature or lower. Specifications vary for different parts of the country and for different periods of the year.

Conduction The transfer of heat by contact. Air is a poor conductor of heat, therefore molecular heat transfer (conduction) affects only two or three feet of air directly. Wind and turbulence, however, continuously bring fresh air into contact with the surface and distribute the warmed or cooled air throughout the atmosphere.

Convection Vertical air motion. The horizontal air movement that completes an air current is called advection.

Cooling processes A method by which air temperature is decreased. The following are all cooling processes: **1. nocturnal cooling**: The Earth continuously radiates its heat outward toward space. During the night the loss of radiant energy lowers the temperature if the Earth's surface. The air temperature is therefore reduced by conduction. **2. advection cooling**: When the windflow is such that cold air moves into an area previously occupied by warmer air, the temperature of the air over the area is decreases. Also, warm air advection over a colder surface will result in conductive cooling of the lower air layers. **3. evaporative cooling**: When rain or drizzle falls from clouds, the evaporation of the water drops cools the air through which these drops are falling. **4. adiabatic cooling**: The process by which air cools due to a decrease in pressure. If air is forced upward in the atmosphere, the resulting decrease in atmospheric pressure surrounding the rising air allows the air to expand and cool. Weather produced by lifting processes, such as frontal weather, convective and orographic thunderstorm, and upslope fog are the result of adiabatic cooling.

Depression A cyclonic or low-pressure, area.

Diurnal Of or referring to actions completed within 24 hours, or pertaining to day time.

Equinox The moment, occurring twice each year, when the Sun crosses the celestial equator. It is so called because the night is equal to the day, each being 12 hours long over the whole Earth. The autumnal equinox occurs on or about September 22 when the Sun is traveling southward; the vernal equinox about March 21, when the Sun is moving northward.

Gradient The rate of increase or decrease in magnitude such as a pressure or temperature gradient. When a horizontal pressure gradient exists, the direct force exerted by the area of higher pressure is called the pressure gradient force. When used to describe a wind (gradient wind), gradient refers to winds above the influence of surface friction — normally above 2000 or 3000 feet — where only pressure gradient force is affecting the speed of the wind.

Greenhouse effect A term that refers to the gradual warming of the Earth caused by material suspended in the atmosphere and reflecting heat back to the Earth's surface instead of allowing the heat to radiate off into space.

Gust A rapid change in wind speed with a variation of 10 knots or more between peaks and lulls.

Horse latitudes The subtropical high-pressure region at about 30 degrees latitude, with calm, or light, variable winds.

Hot wave or heat wave A period of unusually high temperatures lasting three or more consecutive days during each of which the maximum temperature is 90°F or over.

Humidity A general term for the water vapor content of the air.

Lapse rate A change in value expressed as a ratio, generally used with temperature changes vertically; (i.e.), 2°C per 1000 feet in the standard atmosphere.

Mean sca level The average height of the surface of the sea for all stages of the tide during a 10-year period.

Mesometeorology The study of atmospheric phenomena such as tornadoes and thunderstorms that occur between meteorological stations or beyond the range of normal observation from a single point.

Micrometeorology The study of variations in meteorological conditions over very small areas, such as hillsides, forests, river basins, or individual cities. Also called microclimates.

Natural air Air as found in the atmosphere, containing water vapor and other impurities.

Radiation Electromagnetic waves traveling at over 186,000 miles per second, many of which are visible as light. Cosmic rays, gamma rays, X-rays, ultraviolet rays, visible light rays, infrared rays, and radio waves are some common types of radiation.

Rate of evaporation The time it takes for a substance to change from a liquid to a gas. The quantity of vapor that will escape from a liquid surface into the air depends on the temperature of the liquid, the amount of vapor already in the air, and the speed of air movement over the liquid surface. Thus, much water will evaporate from the Great Lakes into the cold, dry winter air above it, even through the air cannot support the moisture in the vapor state. The resulting condensation forms a dense evaporation fog over the lakes.

Solstice The time of year when the direct ray of the Sun is farthest from the Equator. The typical dates are: summer solstice, June 22; winter solstice, December 22.

Spread The difference between the temperature of the air and the dew point of the air, expressed in degrees. Although there is a definite relationship between spread and relative humidity, a spread of 5°F between 90°F and 85°F produces a significantly different relative humidity from the same spread between 65°F and 60°F.

Squall A line of thunderstorms, generally continuous, across the horizon. The squall line is associated with prefrontal activity.

Syntopic That which presents a general view of the whole; as a syntopic weather map or a syntopic weather situation in which the major weather phenomena over a large geographical area are depicted or discussed.

Temperature lag Although the sun is directly overhead at noon, incoming radiation continues to exceed reradiation from the Earth until after 2 p.m. local standard time. Thus, the surface temperature increase reaches a maximum in the mid-afternoon. Also, seasonally, the sun is highest in the Northern Hemisphere at the summer solstice (June 22), but the long hours of daylight and relatively direct incident radiation cause the summer temperatures to continue increasing into July and August.

Weather The state of the six meteorological elements in the atmosphere: air temperature, humidity, clouds, precipitation, and wind. The term weather also has several specialized meanings in the atmosphere at the time of a meteorological observation, both the precipitation and obstructions to vision (fog, haze, smoke, dust, etc.), or all forms of atmospheric phenomena.

Wind velocity The speed and direction of the wind.

Index

A

advection, 44
advection fog, 131
agriculture, weather and, 127-130
agrometeorology, 127-130
air masses, 15, 19-38
  age of, 25-26
  arctic-antarctic, 19, 21, 22
  changes in, factors affecting, 23-26
  classification of, 19
  clouds and, 20
  cold, 20, 27
  continental, 19, 22, 24, 115-117
  continental tropical, 117
  equatorial, 19, 22
  fronts and (see fronts)
  heating and cooling of, 26
  lifting of, 26
  lows and highs in (see low pressure; high pressure)
  maritime, 19, 22, 25, 113-116
  maritime polar, 113-115
  mixing of, 26
  moisture in, 24-26
  monsoon, 19, 20
  mountains and, 25
  path of, 25
  polar, 19, 22, 24, 25, 113-116
  sinking of, 26
  source regions for, 21-23
  surface conditions that change, 23-25
  thermal and mechanical modifications to, 26
  tropical, 19, 22, 116, 117
  warm, 20, 27
  wintertime, North America, 115
air travel, weather and, 130-134
altitude, micro-climates and, 123-124
altocumulus, 29, 59, 96
altostratus, 27, 29, 30, 59, 99
anemometer, 80
aneroid barometer, 83
animal behavior, weather forecasts based on, 70, 71, 74
Antarctic Circle, 7
anticyclones, 37
Arctic Circle, 7
arctic fronts, 26
arctic-antarctic air masses, 19, 21, 22
area forecast, 106
atmosphere, 1-17
  circulation of, 14, 15, 23, 50, 118
  composition of gases in, 11
  coriolis force, 50
  layers of, 9-13
  pressure of, 45-50
  three-cell theory of circulation in, 118

B

barometer, 80-83
biometeorology, 72-73
boating, weather and, 134-136, 134
boiling point, 39
Brandes, Heinrich, 67
business, weather and, 136

C

Canadian Meteorological Service, 67
Celsius, 39
centigrade, 39
chromosphere, 3, 4
cirrocumulus, 59, 96, 120
cirrostratus, 27, 30, 59, 96
cirrus, 29, 30, 59, 71, 96, 119
clamatology, 2
climate, 2
  causes for variations in, 13-16
  distribution of land and water and, 14-16, 42, 43, 123
  latitude and, 13
  length of days and, 13
  micro-, 123-124
  prevailing elements of, 15
  topography and, 15
climatological forecasting, 122
climatology, 123
clinometer, 91
clouds, 27, 29, 30, 33, 57-61, 71, 73, 74, 96, 99, 107, 109, 119, 120, 130
  air masses and, 20
  ceilings of, 90, 91, 132
  classification, abbreviation, symbols for, 59
  cumuliform, 58
  flying and, 132
  fracto, 59
  funnel (see tornadoes; waterspouts)
  high, middle, low, 58
  hydrologic cycle, 61
  measuring, clinometer for, 90-92
  nimbus, 59
  obstruction of, 90
  stratiform, 58
  symbols for, 94
  visibility and, 91, 92
cold air masses, 20, 27
cold fronts, 27-29
  fast-moving, 29
  frontal slope of, 28
  slow-moving, 27-29
cols, 37
condensation, 11, 25, 53, 61
conduction, 44
continental air masses, 19, 22, 24, 115-117
continental polar air masses, weather associated with, 115-116
continental tropical air masses, weather associated with, 117
convection, 44
convective currents, 60
convergence, 60
cooling degree days, 136
Coriolis force, 50
corona, 3, 4
cumuliform, 33, 58
cumulonimbus, 27, 29, 59, 60, 99, 132

144

INDEX

cumulus, 29, 60, 99
cyclone, 35, 37
   wave, life cycle of, 36

D
dew point, 24, 55, 107, 109
drizzle (see also rain), 56, 95, 131

E
Earth, 5-8
   air circulation around, 14, 15, 23, 50, 118
   atmospheric layers of, 9-13
   Coriolis force, 50
   distribution of land and water on, 14-16, 42, 43, 123
   revolution of, 5, 42
   rotation of, 5, 41
   solstices and equinoxes of, 5, 6, 7
   Sun's energy intercepted by, 2, 6, 42, 43, 44, 123
   tilt of, 6, 14, 42
   topography of, 15
   zones of, 7, 8
effective temperature, 136, 137
elements, 39-65
equatorial air masses, 19, 22
Equatorial Zone, 7
equinoxes, 5, 6, 7
evaporation, 25, 53, 61
exosphere, 9, 13

F
Fahrenheit, 39
fan psychrometer, 90
farming, weather and, 127-130
flying, weather and, 130-134
fog, 30, 34, 53, 74, 95, 130, 131
forecasting, 2, 67, 77, 118-122
   climatological, 122
   improving accuracy of, 125
   local, state, zonal, special, 99
   numerical, 122
   outlooks, 101
   statistical, 121
   syntopic, 121
   warnings, 99
   watches, 99
fractocumulus, 59
frontal lift, 60
frontal zone, 26
fronts, 15, 26-35, 119
   arctic, 26
   cold, 27-29
   occluded, 30, 119
   polar, 22, 26, 27
   stationary, 32-35
   vertical view of, 27
   warm, 28-30
frost, 74
funnel clouds, 60

G
gales, 135
gamma rays, 8
generalized physical regions, U.S., 3
gradient, wind, 50
gradients, pressure, 49, 50
greenhouse effect, 43
ground fog, 131
gusts, 79

H
hail, 53, 57, 94, 130
halo, Sun or Moon, 71
health, weather and, 72-73
heat (see temperature)
heat exchange, 55
heat transfer, 43, 44
heating degree days, 136
Henry, Joseph, 67
high pressure systems, 35-38, 49, 50
   air circulation around, 36, 37, 38
high-pressure area, 21
humidity, 95
   measuring, hygrometers for, 85-89
   specific and relative, 55-57, 86, 95, 124
hurricanes, 35, 135
hydrologic cycle, 61-62
hygrometers, 85-89

I
ice, 33, 34, 53, 57, 95, 132
   flying and, 134
inches of mercury, 48
industrial meteorology, 136
infrared rays, 8
insolation, 8, 43
isallobars, 108
isobars, 33, 48, 108

J
jet streams, 52-54

L
land, distribution of continents, 14-16, 42, 43, 123
lapse rate, temperature, 40, 41
latitude, 13
light (see clouds; Sun)
log book, 93-99
low ceilings, 90
low pressure systems, 35-38, 49, 50
   air circulation around, 36, 37, 38

M
man, weather's effect upon, 72-73
marine weather advisory signals, 135
maritime air masses, 19, 22, 25, 113-116
maritime polar air masses, weather associated with, 113-115
maritime transport, weather and, 134-136
maritime tropical air masses, weather associated with
maximum-minimum thermometers, 85
melting point, 39
mercury barometers, 82
mesosphere, 13
meteorology, 1, 67
micro-climates, 123-124
millibars, 48
moisture (see also humidity; precipitation), 23, 24, 25, 53-58
   condensation, 53
   dew point, 55
   evaporation, 53

heat exchange and, 55
hydrologic cycle, 61
precipitation, 55
saturation, 55
specific and relative humidity, 55, 56, 57
sublimation, 53
monsoon air masses, 19, 20
Moon, halo around, 71
mountains, 25
   wind and, 53

N
National Meteorological Center (NMC), 68, 106
National Oceanic-Atmospheric Administration (NOAA), 67, 103
National Weather Service, 67, 68, 99-101m 121-122
   boating weather and, 134
   distribution of information from, 101-106
   forecasts from, 99
   outlooks, 101
   raw weather data from, 101
   watches and warnings, 99
newspaper weather forecast page, 102
nimbostratus, 27, 30, 59, 99
nimbus clouds, 59
nitrogen, 11
numerical forecasting, 122

O
occluded fronts, 30-33, 119
   cold, 31
   warm, 31
oceans, 14-16, 42, 43, 123
orographic lift, 60
outlooks, 101
oxygen, 11
ozone layer, 12

P
peak gusts, 79
photosphere, 3, 4
pilots, weather and, 130-134
plages, solar, 4
polar air masses, 19, 22, 24, 25, 113-116
polar fronts, 22, 26, 27
polar zones, 8
pores, solar, 4
precipitation, 15, 25, 55, 56, 73, 94, 96, 109, 119
   animal behavior to forecast, 70, 71, 74
   fog and, 131
   hydrologic cycle, 61
   measuring, rain gauges for, 89-90
   record highes, worldwide, 65
   record lows, worldwide, 65
   world pattern of, 58
pressure, 45-50, 72, 95, 107, 109, 117, 118, 119
   forecasting chart using wind and, 120
   isobars, 48, 108
   lows, highs, pressure gradients, 49, 50

measurement of, 48
measuring, barometers for, 80-83
movement of, 107, 109
normal at sea level, 48
prevailing systems in January, worldwide, 121
prevailing systems in July, worldwide, 122
symbols for, 94
wind and, 50
wind flow and, 114
prevailing elements, 15
prevailing westerlies, 117
proverbs about weather, 68-72
psychrometer, 85, 86, 87, 90

R
radiation, 43
radiation fog, 131
radio weather forecasts, 103
rain, 27, 29, 30, 33, 34, 53, 56, 57, 60, 68-71, 94, 131
rain gauges, 89-90
reflection, 8, 9
relative humidity, 55, 56, 57, 86, 95
   micro-climates and, 124
   temperature vs., 56, 137
ridge, 37
rotor psychrometer, 88, 89

S
sailing, weather and, 134-136
saturation, 55
sayings about weather, 68-72
seasons, 5, 6, 7
six's type thermometers, 84
sky observations, 73-74, 107
   blue skies, 74
   storm indicators, 74
   symbols for, 94
sleet, 53, 57, 94
sling psychrometer, 88
small craft advisories, 134, 135
snow, 27, 30, 33, 53, 94, 130
solar flares, 4, 5
solar prominences, 3, 5
solar radiation, 8-9
   cooking with, 9
   electricity from, 8
solstices, 5, 7
specific humidity, 55, 56, 57, 86, 95
squall lines, 29, 132
squalls, 79
station models, 107-109
stationary fronts, 32-35
   vertical cross section of, 33-35
statistical forecasting, 121
storms (see also thunderstorms), 135
   indicators of, 74
stratiform, 33, 57, 58
stratocumulus, 59, 99
stratopause, 12
stratosphere, 12
stratus, 30, 59, 99
stream fog, 131
sublimation, 53, 57
Sun, 2-5
   flares, prominences, sunspots on, 3, 4, 5

Sun (continued)
  halo around, 71
  layers of, 3, 4
  solar radiation from, 8-9
  sunshine, mean annual total hours of, 41
  temperature of, 2
  thermonuclear reactions of, 3
sunspots, 4
superior air masses, 19
surface syntopic weather map, 106
symbols
  clouds, 94
  pressure, 94
  sky conditions, 94
  weather conditions, 94
  weather map, 110-112
syntoptic forecasting, 121

T
teletype weather forecasts, 103
Temperate Zones, 7
temperature, 15, 23, 39-45, 95, 107, 109
  advection, 44
  altitude vs., 12, 40, 41
  average January, worldwide, 129
  average July, worldwide, 129
  changes in, factors affecting, 41-45
  conduction, 44
  convection, 44
  converting between Celsius and Fahrenheit, 39-40
  cooling degree days, 136
  effective, 136, 137
  greenhouse effect, 43
  heat transfer and, 43, 44
  heating degree days, 136
  highest by month, U.S., 45
  insolation, 43
  lapse rate, 40, 41
  lowest by month, U.S., 45
  measuring, thermometers for, 83-85
  micro-climates and, 123-124

  normal daily maximum for July, 42
  normal daily minimum for January, 43
  radiation, 43
  record highes, by State, 46
  record highes, worldwide, 65
  record lows, by State, 47
  record lows, worldwide, 65
  relative humidity vs., 56, 137
  wind chill, 138, 139
temperature-humidity index (THI), 137
terminal forecast, 106
thermal lows, 35
thermometers, 83-85
thermosphere, 13
three-cell theory of air circulation, 118
thunderstorms, 23, 27, 29, 30, 34, 60-62, 72, 94, 130, 132, 133, 135
time conversion table, 100
topography, 15, 23
tornadoes, 35, 60-61
  average monthly days in U.S. for, 64
  mean monthly average, by state, 63
Torrid Zone, 7
trade winds, 23
transmissometer, 92
Tropic of Cancer, 7
Tropic of Capricorn, 7
tropical air masses, 19, 22, 116, 117
tropical storms, 35
Tropical Zone, 7
tropopause, 11
troposphere, 10-12
troughs, 35, 37
turbulence, 130, 132
typhoons, 35

U
ultraviolet rays, 8, 12
unlimited ceiling, 91

V
visibility, 91, 130, 132
  measuring, transmissometer, 92

W
wake turbulence, 132
warm air masses, 20, 27
warm fronts, 29-30
  frontal slope of, 28
  vertical cross section of, 31
warnings, 99
watches, 99
water, distribution of oceans, 14-16, 43, 43, 123
water vapor, 11, 24
waterspouts, 35, 60-61
wave cyclone, 36
weather, 2
  comfort and, 136
  health effects of, 72-73
  man vs., 1-2
  proverbs about, 68-72
  recording observations (see weather data collection)
  sky observations, 73-74, 77
  symbols for, 94
weather data collection, 74, 75, 93-112
  distribution of weather information for, 101-106
  National Weather Service and, 99-101
  newspaper forecast, 102
  station model, 107-109
  teletype, 103
  weather log for, 93-99
  weather maps, reading, 94, 106-109
  Weather Radio broadcasts, 103
  wind rose, 96
weather data interpretation, 113-126
  climatology and, 123
  continental polar air masses, weather from, 115-116
  continental tropical air masses, weather from, 117

  improving forecast accuracy, 125
  maritime polar air masses, weather from, 113-115
  maritime tropical air masses, weather from, 116-117
  micro-climates and, 123-124
  plotting weather for, 117, 117
weather log, 93-99
weather maps, 106-109
  symbols used on, 110-112
Weather Radio, 103
  stations and their frequencies, 104-105
Weather Service Forecast Offices, 68
weather shack, 78, 79
wet bulb depression, 86
wind, 15, 50-53, 68, 69, 70, 73, 118, 119, 120, 130
  character of, 79
  direction of, 78, 79, 80, 107, 124
  electricity from, 51
  forecasting chart using pressure and, 120
  jet streams and, 52-54
  land-sea flow of, day and night, 52
  measuring, 78
  micro-climates and, 123-124
  mountains and, 53
  pressure and, 50
  prevailing westerlies, 117
  shifts in, 79
  speed of, 79, 80, 107
  speed table, 97-98
  surface flow in pressure systems, 114
  turbulence and, 132
  wind chill, 138, 139
  wind rose, 96
  wind shear, 130, 132
wind chill, 138, 139
wind rose, 96
wind shear, 130, 132
wind vanes, 79, 80
World Meteorological Organization (WMO), 68